AF808655

***FM 90-4**

FIELD MANUAL
No. 90-4

HEADQUARTERS
DEPARTMENT OF THE ARMY
Washington, DC, 16 March 1987

AIR ASSAULT OPERATIONS

Contents

Page

Preface iii

CHAPTER 1 Air Assault Operations in the AirLand Battle
Section I AVIATION AND INFANTRY 1-1
II ORGANIZATION OF AIR ASSAULT FORCES 1-2
III CAPABILITIES, LIMITATIONS, AND VULNERABILITIES 1-2
IV EMPLOYMENT 1-4

CHAPTER 2 Preparation for Combat
Section I PROCEDURES 2-1
II INTELLIGENCE PREPARATION OF THE BATTLEFIELD 2-1
III THE THREAT 2-2
IV TASK ORGANIZING FOR AIR ASSAULT OPERATIONS 2-3
V COMMAND, CONTROL, AND COMMUNICATIONS 2-9

CHAPTER 3 Operations Planning
Section I ESTIMATE PROCESS 3-1
II SECURITY AND CONTROL 3-6
III GROUND TACTICAL PLAN 3-7
IV THE LANDING PLAN 3-8
V AIR MOVEMENT PLAN 3-11
VI LOADING PLAN 3-26
VII STAGING PLAN 3-36
VIII AIR MISSION BRIEFING 3-36

CHAPTER 4 Combat Operations
Section I OFFENSE 4-1
II DEFENSE 4-10
III OTHER TACTICAL MISSIONS 4-12

DISTRIBUTION RESTRICTION: Approved for public release, distribution is unlimited.
*This publication supersedes FM 90-4, 8 October 1980.

CHAPTER 5 Combat Support
Section I COMBINED ARMS TEAM 5-1
II FIRE SUPPORT 5-2
III FIRE SUPPORT PLANNING AND COORDINATION 5-3
IV ARTILLERY AIR ASSAULT OPERATIONS 5-8
V NAVAL GUNFIRE SUPPORT 5-10
VI UNITED STATES AIR FORCE TACTICAL AIR SUPPORT 5-10
VII AIR DEFENSE 5-11
VIII ENGINEER SUPPORT 5-12

CHAPTER 6 Combat Service Support
Section I PLANNING 6-1
II COMBAT SERVICE SUPPORT ORGANIZATION 6-2
III COMBAT SERVICE SUPPORT EXECUTION 6-5
IV AVIATION SUPPORT CONSIDERATIONS 6-7

APPENDIX A Conduct of an Air Assault Operation A-1

APPENDIX B Air Assault Planning Forms B-1

APPENDIX C Sample Annex to Infantry Battalion SOP for Air Assault Task Force Operations C-1

APPENDIX D Helicopter Characteristics D-1

APPENDIX E Pickup Zone and Landing Zone Operations E-1

APPENDIX F Air Assault Training F-1

APPENDIX G Army Airspace Management G-1

APPENDIX H Moving Vehicle-mounted TOW by Helicopter H-1

GLOSSARY Glossary-1

REFERENCES References-1

Preface

This manual describes how infantry and aviation units plan and conduct air assault operations. It emphasizes the coordination necessary between these organizations concerning the planning sequence and tactical employment of both elements. It is written primarily for aviation and infantry units and is applicable to combat support and service support units with a need to plan for and use Army aviation support.

Air assault operations are conducted with speed, secrecy, and precision by a well trained, proficient combined arms team. To gain proficiency, individuals and units habitually conduct combined arms training in air assault operations before being committed to combat.

The proponent of this publication is HQ TRADOC. Submit changes for improving this publication on DA Form 2028 (Recommended Changes to Publications and Blank Forms) and forward it to: Commandant, US Army Infantry School, ATTN: ATSH-B-ID, Fort Benning, GA 31905-5593.

The provisions of this publication are the subject of the following international agreements: 2351, Procedures for Marshalling Helicopters in Airmobile Operations; 2860, Principles of Engagement for the Landing Sites; 2861, Procedures for the Recovery of Downed Aircraft/Helicopters While Engaged in Airmobile Operations; 2863, Minimum Navigational Facilities for Multi-National Airmobile Operations; 2876, Planning and Coordination Procedures for Airmobile Operations; 2904, Airmobile Operations (ATP 41); 3117, Aircraft Marshalling Signals; 3345, Data Forms for Planning Air Movements; 3468, General Rues Covering the Transport of Loads by Helicopter; 3532, Transport of Troops by Helicopter; 3570, Drop Zones and Extraction Zones; 3597, Helicopter Tactical or Nonpermanent Landing Sites; 3619, Helipad Markings; 3627, Helicopter Day and Night Formation Flying; and 3630, Helicopter Tactical Operations at the High Hover.

Unless otherwise stated, whenever the masculine gender is used, both men and women are included.

CHAPTER 1

Air Assault Operations in the AirLand Battle

Section I

AVIATION AND INFANTRY

1-1. General.

Army aviation and infantry units can be fully integrated with other members of the combined arms team to form powerful and flexible air assault task forces that can project combat power throughout the entire depth, width, and breadth of the modern battlefield with little regard for terrain barriers. The unique versatility and strength of an air assault task force is achieved by combining the capabilities of modern rotary-wing aircraft - speed, agility, and firepower - with those of the infantry and other combat arms to form tactically tailored air assault task forces that can be employed in low-, mid-, and high-intensity environments.

1-2. Control.

Air assault operations are those in which assault forces (combat, combat support, and combat service support), using the firepower, mobility, and total integration of helicopter assets, maneuver on the battlefield under the control of the ground or air maneuver commander to engage and destroy enemy forces or to seize and hold key terrain. Air assault operations are not merely movements of soldiers, weapons, and materiel by Army aviation units and must not be construed as such. They are deliberate, precisely planned, and vigorously executed combat operations designed to allow friendly forces to strike over extended distances and terrain barriers to attack the enemy when and where he is most vulnerable.

> **NOTE:** Air movement operations are those operations involving the use of Army airlift assets for other than air assaults. These operations are used to move troops and equipment, to emplace artillery pieces and air defense artillery (ADA) systems, and to transport amrnunition, fuel, and supplies. The same general plans used for air assault operations may need to be prepared for large-scale air movement operations. In these operations, aviation is not task-organized with other members of the combined arms team to engage enemy forces. When an airlift is completed, the air movement operation is terminated and, unless otherwise specified in the order, aviation units are released to return to their parent units.

1-3. Commanders.

To take advantage of the opportunities offered by an air assault task force, commanders and leaders must develop an insight into the principles governing their development (organization) and employment.

1-4. Infantry.

Although air assault, airborne, ranger, and light infantry units are much more suited to the role than are other types of infantry, all infantrymen and their supporting arms counterparts must be prepared to execute air assault operations when the situation dictates. Mechanized infantry units of the heavy division can exploit the mobility and speed of organic or supporting helicopters to secure a deep objective in the offense, reinforce a threatened sector in the defense, or to place combat power at a decisive point on the battlefield. For this reason, they must be proficient in the conduct of air assault operations.

Section II

ORGANIZATION OF AIR ASSAULT FORCES

1-5. General.

There are no existing units below division level that are capable of unilaterally conducting effective air assault operations. Pure units simply do not have adequate organic assets to ensure successful air assault mission accomplishment. Task organizing or mission-specific tailoring of forces is the norm for air assault operations.

1-6. Task force.

Air assault operations are accomplished by employing an air assault task force (AATF). The AATF is a group of integrated forces tailored to the specific mission and under the command of a single headquarters. It may include some or all elements of the combined arms team. The ground or air maneuver commander, designated as the air assault task force commander (AATFC), commands the AATF. The AATFC may combine infantry companies with aviation assets that can be employed singly or in multiples. (For a discussion of how AATFs are organized, see Chapter 2.)

Section III

CAPABILITIES, LIMITATIONS, AND VULNERABILITIES

1-7. General.

An air assault task force provides commanders with truly unique capabilities. They can extend the battlefield, move, and rapidly concentrate combat power like no other available forces.

1-8. Capabilities.

Specifically, an air assault task force can:

a. Attack enemy positions from any direction.

b. Delay a much larger force without becoming decisively engaged.

c. Overfly or bypass barriers and obstacles and strike objectives in otherwise inaccessible areas.

d. Conduct deep attacks and raids beyond the forward line of own troops (FLOT) or line of contact (LC), using helicopters to insert and extract forces.

e. Rapidly concentrate, disperse, or redeploy to extend the area of influence.

f. Provide responsive reserves allowing commanders) to commit a larger portion of his force to action.

g. React rapidly to tactical opportunities and necessities; conduct exploitation and pursuit operations.

h. Rapidly place forces at tactically decisive points in the battle area.

i. Provide surveillance or screen over a wide area.

j. React to rear area threats.

k. Rapidly secure and defend key terrain (such as crossing sites, road junctions, bridges) or deep objectives.

l. Bypass enemy positions; achieve surprise.

m. Conduct operations under adverse weather conditions and at night to facilitate deception and surprise.

n. Conduct fast-paced operations over extended distances.

o. Conduct economy-of-force operations over a wide area.

p. Rapidly reinforce committed units.

1-9. Limitations.

An air assault task force is light, mobile, and relies on helicopter support throughout any air assault operation. As such, they may be limited by:

a. Adverse weather, extreme heat and cold, and other environmental conditions such as blowing snow and sand that limit flight operations or helicopter lifting capability.

b. Reliance on air lines of communication.

c. Hostile aircraft, air defense, and electronic warfare action.

d. Reduced ground mobility once inserted.

e. Availability of suitable landing zones (LZ) and pickup zones (PZ).

f. Available nuclear; biological, chemical (NBC) protection and decontamination capability.

g. Reduced vehicle-mounted antitank weapon systems (except in air assault units).

h. Battlefield obscuration that limits helicopter flight.

i. High fuel (JP4) and ammunition consumption rates.

1-10. Vulnerabilities.

An air assault task force uses the helicopter to move to and close with the enemy. Initial assault elements must be light and mobile. They are often separated from weapon systems, equipment, and materiel that provide protection and survivability on the battlefield. Thus, an air assault task force is particularly vulnerable to enemy:

a. Attack by aircraft and air defense weapon systems during the movement phase.

b. Attack by NBC systems, because of limited NBC protection and decontamination.

c. Attacks (ground, air, or artillery) during the loading and unloading phases and at other times when the infantry is not dug in.

d. Air strikes, due to limited availability of ADA weapon systems that can be deployed with an air assault task force.

e. Electronic warfare (jamming), due to the heavy reliance on radio communications for command and control (C2).

f. Artillery or other fires that may destroy helicopters and air assault forces during PZ or LZ operations.

g. Small arms fire that presents a large threat to helicopters.

Section IV

EMPLOYMENT

1-11. General.

Air assault operations are high risk, high payoff operations, that, when properly planned and vigorously executed, allow commanders to apply the four basic tenets and 10 combat imperatives of the AirLand Battle Doctrine (FM 100-5). An air assault task force can dramatically extend a commander's area of operation, enabling him to execute AirLand Battle Doctrine in areas ranging beyond the capability of more conventional forces.

1-12. Tactical employment.

The tactical employment of an air assault task force is different from those of light and other dismounted infantry. An air assault task force is employed judiciously and only on missions that require:

- Massing or shifting combat power rapidly.
- Using surprise.

- Using flexibility, mobility, and speed.
- Gaining and maintaining the initiative.
- Extending the depth, width, or breadth of the battlefield.

1-13. Operational guidelines.

An air assault task force is normally a highly tailored force specifically designed to hit fast and hard. They are best employed in situations that provide the air assault task force a calculated advantage due to surprise, terrain, threat, or mobility. The principles of employment are basic guidelines that govern the planning and execution of air assault operations. They are:

a. The air assault task force should normally be assigned only missions that take advantage of their superior mobility and should not be employed in roles requiring deliberate operations over an extended period of time.

h. The air assault forces always fight as a combined arms team.

c. The availability of critical aviation assets is a major factor in any operation.

d. The air assault planning must be centralized and precise; execution must be aggressive and decentralized.

e. The air assault operations may be conducted at night or during adverse weather, but require more planning and preparation time in those cases.

f. Unit tactical integrity must be maintained throughout an air assault. When planning loads, squads are normally loaded intact on the same helicopter, with platoons located in the same serial. This ensures fighting unit integrity upon landing.

g. The fire support planning must provide for suppressive fires along flight routes and in the vicinity of landing zones. Priority for fires must be to the suppression of enemy air defense systems (SEAD).

h. Infantry unit operations are not fundamentally changed by integrating aviation units with infantry; tempo and distance are dramatically changed, however.

i. Although mechanized infantry units are not frequently employed in air assault operations, such operations conducted on a limited scale may be the decisive form of combat. Typical air assault operations conducted by mechanized forces are river-crossing operations, seizure of key terrain, raids, and rear area combat operations.

j . An air assault task force is employed most effectively in environments where limited lines of communication are available to the enemy, where he lacks air superiority and effective air defense systems.

CHAPTER 2

Preparation for Combat

Section I

PROCEDURES

2-1. General.

The AATFC must prepare for air assault operations by following troop leading procedures and organizing for a specific mission. This chapter discusses procedures and organization, providing a basis for detailed discussion of air assault operations in later chapters.

2-2. Procedures.

The following sections discuss combat preparation procedures:

- Intelligence preparation of the battlefield.
- The Threat.
- Task organizing for air assault operations.
- Command, control, and communications.

Section II

INTELLIGENCE PREPARATION OF THE BATTLEFIELD

2-3. General.

Intelligence preparation of the battlefield (IPB) is a systematic approach to analyzing the enemy, weather, and terrain in a specific geographic area. It integrates enemy doctrine with the weather and terrain as they relate to the mission and the specific battlefield environment. This is done to determine and evaluate enemy capabilities, vulnerabilities, and probable courses of action. The main thrust of IPB is to support commanders and their staffs in the decision-making process. It results in a graphic intelligence estimate that portrays probable enemy courses of action. Once hostilities begin, and current data becomes available, the IPB intelligence estimate becomes dynamic, changing with the immediate situation on the battlefield.

2-4. Intelligence preparation of the battlefield.

Intelligence preparation of the battlefield is a sequential process of intelligence analysis that orients on the assigned areas of operations and interest, and the enemy forces that are expected to be operating in those areas. The five logical steps include Threat evaluation, areas of operation and interest evaluation, terrain analysis, weather analysis, and Threat integration.

2-5. Graphics.

The use of graphics is key to IPB. Threat evaluation and Threat integration are accomplished through the analytical techniques known as templating. A template is a graphic illustration of enemy force structure, deployment, or capabilities normally drawn to scale. It provides a basis for command judgment and decisions affecting resource allocation. It is used as a comparative data base to integrate what is known about the enemy with a specific weather and terrain scenario. Templates enable planners to visualize enemy capabilities, predict likely courses of action before the battle, and confirm or refute them during combat. The four principal templates are developed during the IPB process:

a. Doctrine. Enemy doctrinal deployment for various types of operations without constraints imposed by weather and terrain. Composition, formations, frontages, depths, equipment numbers and ratios, and high value targets (HVT) are types of information displayed.

b. Situation. Depicts how the enemy might deploy and operate within the constraints imposed by the weather and terrain.

c. Event. Depicts locations where critical events and activities are expected to occur and where critical targets will appear.

d. Decision points. Depicts decision points keyed to significant events and activities; the intelligence estimate in graphic form.

2-6. Weather.

Planners must not underplay the effects of weather on air assault operations. It has a significant impact on both friendly and enemy air capabilities. The temperature humidity combination affects helicopter lift capability because of density altitude conditions. Weather factors also affect conditions of LZs, air avenues of approach, and Threat air defense weapons.

2-7. Intelligence preparation of the battlefield products.

These are routinely used by air assault task forces throughout the planning phases. Air assault operations are high-risk operations at best, and they should be planned with the best possible intelligence support available. Brigades and battalions do IPB on an informal basis as time and resources permit. Corps and division G28 must be prepared to provide detailed IPB support to any subordinate unit that has been assigned an air assault mission.

Section III

THE THREAT

2-8. General.

The primary Threat tactics against air assault operations can be broken down into four major areas:

- Air defense fires (including small arms).
- Fixed- and rotary-wing aircraft.
- Electronic warfare.
- Threat reaction to landing zone operations.

a. Vulnerability to air defense fires must be recognized and compensated for by effective suppressive measures and increased emphasis on accurate, timely, intelligence of the enemy.

b. The capabilities and limitations of Threat aircraft within the area of operation must be understood and all measures to minimize the risk of encounter must be taken.

c. Threat electronic warfare capabilities that would influence the air assault operation to include: jamming, direction finding and monitoring of communications, or jamming and direction finding involving friendly radars must be considered and appropriate electronic countermeasures employed.

d. Analysis of Threat capabilities to interdict friendly landing zones with ground forces, artillery, and close air support must be accomplished during the planning phase of the operation.

2-9. Intelligence preparation of the battlefield exploits weaknesses.

A major portion of the IPB is understanding the enemy. Knowledge of enemy doctrine, tactics, and equipment enables an air assault task force to find and exploit weak points.

Section IV

TASK ORGANIZING FOR AIR ASSAULT OPERATIONS

2-10. General.

As stated in Chapter 1, air assault operations are not conducted by pure units, but rather by tactically tailored AATFs (brigade or battalion level) designed to accomplish a specific mission. Organizing the task force for combat is a significant action. Predesignated and well-understood command and support relationships ensure that the force will fight as a cohesive, coordinated team. Normally:

a. The formation of an AATF will be directed by a headquarters no lower than division level (or that which can allocate dedicated aviation resources).

b. The directing or establishing headquarters allocates assets, defines authority and responsibility by designating command and support relationships, and forms the AATF early in the planning stage. Divisional aviation assets in other than the air assault division may be inadequate; therefore, additional aviation resources must be requested from corps units.

c. Battalion is the lowest level staffed with sufficient personnel to plan, coordinate, and control an air assault operation. When company-sized operations are conducted, the predominance of planning occurs at battalion or higher level.

d. An AATF will exist only until completion of a specified mission. After that, aviation and other elements are returned to the control of their parent unit(s).

2-11. Considerations for developing an air assault task force.

a. The availability of aviation assets is normally the major factor in determining AATF task organization.

b. The AATF must provide a mission-specific balance of mobility, combat power, and staying (sustaining) power.

c. The required combat power should be delivered to the objective area as soon as possible, consistent with aircraft and PZ capacities, to provide surprise and shock effect.

d. To perform its mission, an air assault task force must arrive intact at the LZ. The force must be tailored to provide en route security and protection from the PZ, throughout the entire flight route(s), and at the LZ.

e. In addition to the traditional command and support relations, one nonstandard command relationship, attached for movement, is used extensively during air assault operations. Under this relationship, some elements, (field artillery [FA], ADA, military intelligence [MI], engineers) may be attached to maneuver elements for movement only. This relationship facilitates command and control, movement planning, and local security of attached elements. Attachment would be effective from the planning phase until landing in the LZ, link up with parent unit, or as predesignated by standing operating procedure (SOP) or operation order (OPORD).

f. The complete AATF is usually formed during the planning phase.

g. The task organization must be determined and announced early in the planning process. It may be included in the warning order.

h. The AATF is normally organized with sufficient combat power to seize initial objectives and protect LZs, and with sufficient combat service support (CSS) and accompanying supplies to sustain a rapid tempo until follow-on or linkup forces arrive, or until the mission is completed.

i. An effective command and control system must be developed for all air assault operations. The AATFC must bring command and control considerations into play as he develops his task organization.

j . Unit tactical integrity must be maintained throughout an air assault. When planning loads, squads are normally loaded intact on the same helicopter. This ensures unit integrity upon landing.

k. Combat support elements are normally placed in direct support (DS) to the AATF in order to ensure close coordination and continuous, dedicated support throughout an operation.

2-12. The air assault task force.

a. The AATF is a tactically tailored combination of combat, combatsupport(CS), and CSS elements under the command and control of a single headquarters or command group.

(1) **The AATF command group and staff.** The AATFC is normally the infantry brigade or battalion commander whose own unit(s) forms the nucleus or predominance of forces in the AATF. He commands the air assault operation and is responsible for its overall planning and

execution. He controls all units assigned, attached, or under operational control (OPCON) to the AATF, and establishes mission priorities for those units in DS or general support (GS) of the AATF.

(2) **The air mission commander.** The air mission commander (AMC) is designated by the supporting aviation brigade or battalion commander and is subordinate to the AATFC. He controls all Army aviation assets in support of the AATF, ensures that aviation operations are conducted according to the AATFC's directives, serves as the AATFC's advisor on aviation matters, and assists the AATFC with planning.

(3) **Aviation liaison officer.** An aviation liaison officer (LO) should be provided to the AATF from the supporting aviation unit and should be considered a special staff officer. His role is to advise the AATFC on all matters relating to Army aviation and to jointly develop, along with the AATF S3 Air, the detailed plans necessary to support the air assault operation. During the execution phase, he should be available to assist the AATFC or S3 Air in coordinating the employment of aviation assets.

b. The AATFC, the AMC, and their respective battle staffs must consider several air assault unique factors, as well as those of mission, enemy, terrain, troops and time available (METT-T), before deciding on the exact AATF task organization. They include both general and organization specific factors (paragraph 2-13).

c. Figure 2-1 depicts a typical AATF organization built around an air assault infantry battalion nucleus.

TEAM A	TEAM B	TEAM C	TEAM D	TASK FORCE CONTROL
A/1-20	B/1-20	C/1-20	D/1-20	120 AVN CO (DS)
GSR TEAM	STINGER TEAM	STINGER TEAM	MORTARS	A/14 ATTACK HEL BN (OPCON)
STINGER SECTION	ENGINEER SQUAD	ENGINEER SQUAD		A/2-10 AIR RECON (OPCON)
				PLT/116 AVN CO (ASH) (DS)
				B/2-16 FA (DS)
				PLT(-)/299 MI (DS)
				PLT(-)/1-2 ADA (DS)
				PLT(-)/19 EN (DS)
				FSSE (DS)
				SCOUT PLT

(See glossary for acronyms and abbreviations.)

Figure 2-1. Typical AATF organization.

2-13. Organization considerations.

Typical roles, missions, and organization-specific considerations are:

a. **Infantry.** Infantry elements normally form the nucleus of the AATF. Although nonmechanized infantry is better suited for air assault operations, there will be situations where mechanized units accomplish their mission by capitalizing oil the helicopter's mobility.

(1) The disposition of the unit's vehicles is a primary point of consideration. When mechanized

infantry units participate in air assault operations, the armored vehicles can be:

(a) Attached for movement to an assaulting ground element (linkup force).

(b) Left in an assembly area until the assaulting element returns.

(c) Repositioned to provide supporting fires for adjacent units or the air assaulting force.

(2) Other considerations include the following:

(a) Ground mobility is limited once the unit is inserted unless vehicles are provided.

(b) Communication range is limited to that of portable radios.

(c) Range of the scout platoon is limited unless its vehicles are lifted into the objective area.

(d) Antiarmor capability is reduced.

(e) Combat support and combat service support will be austere.

(f) Air lines of communication must be planned for sustainment.

b. **Assault (lift) helicopters.**

(1) **Organization.** The AATF would normally be one or more assault helicopter platoons or companies (depending on the size of the operation) placed under OPCON or in DS to the AATF for the duration of the operation.

(2) **Role.** The assault helicopters operate under the control of the AMC who will direct actions based on the AATFC's order.

(3) **Typical missions.** These include-.

(a) Tactical mobility for troops, equipment, and weapon systems by internal and external load.

(b) Aerial resupply by internal and external load.

(c) Backup medical evacuation (medevac).

c. **Attack helicopters.**

(1) **Organization.** Attack helicopter companies or an entire battalion may operate under OPCON to the AATF.

(2) **Role.** Attack helicopter units are normally employed as air maneuver elements in the antiarmor role; however, during air assault operations, they additionally support the lift and assault force by direct and indirect fires in the absence of normal artillery and other fires.

(3) **Typical missions.** These include:

(a) Protect (escort) lift helicopters from the PZ to LZ as dictated by the enemy.

(b) Suppress enemy ADA and other weapons en route to and during insertions and/or extractions.

(c) Provide preparatory and/or suppressive fires in the vicinity of LZS or objectives in the absence of conventional artillery.

(d) Overwatch the LZ and objective areas to neutralize enemy resistance and to block enemy attempts to reinforce the objective area.

(e) Serve as AATF reserve when facing a motorized or armored enemy.

(f) Provide reconnaissance and security in the absence of mobile ground scouts or air reconnaissance units.

d. **Air reconnaissance.**

(1) **Organization.** The AATF would normally receive OPCON of an air reconnaissance team or troop.

(2) **Role.** Air reconnaissance elements provide reconnaissance and limited security for the AATF during all phases of the operation and fill the void created by the absence of mobile infantry scouts.

(3) **Typical missions.** These include:

(a) Reconnaissance of PZs, flight routes, LZs, and objectives.

(b) Screening forward (or all-round) of ground forces to provide limited security and early warning.

(c) Providing downed aircraft security.

e. **Assault support (medium) helicopters.**

(1) **Organization.** The AATF may be supported by medium helicopter **pla**toons (or company[s]) placed under OPCON or in DS.

(2) **Role.** Medium helicopters normally are employed in follow-on echelons to build combat power and to resupply the AATF.

(3) **Typical missions.** These include moving:

(a) Artillery (up to M198 in size and weight) and ammunition.

(b) Engineer equipment and barrier materials.

(c) Military intelligence assets.

(d) All classes of supply.

(e) Bridging assets.

(f) Nuclear, biological, and chemical defense and decontamination equipment.

(g) Personnel from secure PZ to secure LZ.

f. **Artillery fire support.**

(1) **Organization.** Field artillery batteries (or battalions) that can be moved by cargo helicopter (CH-47), or that can fire into the air assault objective area, are normally attached to or placed in DS of the AATF.

(2) **Role.** Field artillery units in air assault operations must be ready to move quickly and frequently to prepared LZs and objectives and to **suppress** enemy artillery and air defense fires.

(3) **Typical missions.** Air assault support missions expected from FA units include:

(a) Suppression of enemy air defense along flight routes and in the vicinity of LZs.

(b) Landing zone preparation.

(c) Conducting artillery raids.

(d) Delivering the field artillery's family of scatterable mines (FASCAM).

g. **Engineers.**

(1) **Organization.** An engineer platoon would normally be placed in DS of the AATF. In many situations, engineers would be attached to infantry units for movement but would revert to DS when communications with their parent headquarters is reestablished.

(2) **Role.** Engineers in the air assault role must be organized to move with infantry and to provide mobility, countermobility, and survivability construction using light equipment (chain saws, handtools), demolitions, natural resources, and ingenuity. Light engineer equipment, such as small earth movers or backhoes, may be moved by medium lift helicopters.

(3) **Typical missions.** These include:

(a) Construct and improve PZs and LZs.

(b) Construct expedient countermobility obstacles using natural materials and demolitions.

(c) Help the infantry dig in.

(d) Emplace point minefields.

(e) Fight as infantry.

(f) Breach obstacles.

h. **Air defense.**

(1) **Organization.** The AATF normally receives, as DS (or attached), a tactically tailored ADA team or platoon equipped with light, air transportable short-range air defense (SHORAD) systems. Air defense artillery assets must be tailored to place a high reliance on man-portable air defense (MANPAD) systems such as the Stinger missile system and towed Vulcans (if available). Normally, Stinger teams are attached to infantry units for movement.

(2) **Role.** In air assault operations, SHORAD must fly with the lead elements in order to be in place to protect follow-on echelons in the objective area. Stinger teams are best suited for this role.

(3) **Typical missions.** These include:

(a) Provide point defense of high value locations including PZs, LZS, objective areas, helicopter rearm-refuel points, and laager sites.

(b) Provide direct fires for ground defense (Vulcans).

i. **Electronic warfare.**

(1) **Organization.** A tactically tailored MI platoon would normally be in DS to the AATF, if the enemy dictates. The platoon must be equipped with (mobile) collecting, jamming, and radar hardware that can be moved by available helicopters.

(2) **Role.** In cases where electronic warfare (EW) capability is needed, but cannot be supported by mobile equipment, the AATFC should request Quickfix, Guardrail, or other assets from higher headquarters.

(3) **Typical missions.** These include:

(a) Disruption of enemy command, control, communication (C3).

(b) Degrading enemy fire support and air defense radio nets.

(c) Ground surveillance (radar).

(d) Collection of electronic intelligence.

j. **Reserves.** Because of their superior mobility, an air assault task force requires smaller reserves than do other forces. During air assault operations, each subordinate maneuver unit may be given an on-order mission to reinforce or assume another unit's mission, or to revert to the task force reserve.

k. **Combat service support** elements. The AATF may be supported by a dedicated, tactically tailored, forward service support element (FSSE) that provides mission specific support to the task force throughout the air assault operation.

Section V

COMMAND, CONTROL, AND COMMUNICATIONS

2-14. General.

a. Command and control is the process of directing and controlling the activities of military forces in order to attain an objective. An air assault C^2 system includes the procedures, facilities, equipment, and personnel to gather information, make plans, communicate changes, and control all ground and air elements in pursuit of the AATF objective.

b. Since the battlefield over which the AATF operates may be extended well beyond the norm, special considerations must be given to the command and control of air assault operations. An AATF C^2 system must communicate orders, coordinate support, and provide direction to the AATF in spite of great distances, enemy interference, and the potential loss of key facilities and individuals. Above all, this system must function quickly and effectively, thus allowing the AATFC to receive and process information and to make decisions faster than the enemy.

2-15. Command and control planning.

The AATFC must address C^2 requirements early in the planning phase of any operation. He must establish an effective C^2 system which allows him to control diverse, widely dispersed air and ground elements between the initial PZ and the final objective. The C^2 system may be subjected to degraded communications due to the extended distances over which the AATF must operate and/or enemy jamming. Although an effective C^2 system must include provisions for two-way radio communications, the AATFC must develop a plan and a system which allows execution despite degraded radio communications. The key to successful air assault C^2 lies in precise, centralized planning **and** aggressive, decentralized execution. The AATFC ensures successful air assault C^2 by:

a. **Effective Task Organizing.** All assets must be tailored into discrete, task-organized elements each with two-way radio communications, unity of command, clearly defined missions and objectives, and provisions for maintaining unit integrity throughout the operation. An effective task organization, with each subelement having a clearly defined mission, allows the AATFC the flexibility to decentralize execution and ensure mission success despite degraded communications, the fog of battle, or unexpected enemy reaction.

b. **Precise planning.** Air assault operations must be precisely planned and well-briefed before execution so that each subordinate leader knows exactly what is expected of him, knows the commander's intent, and knows he can execute his mission despite the loss of radio communications. Contingencies or alternatives must be built into each plan to allow for continuation of the mission in a fluid environment. Events must be planned to occur based on time (time driven) or the execution of a previous event (event driven) so that actions will occur at the specified time or in the specified sequence despite degraded communications. For example:

(1) A time-driven event might be the firing of a landing zone FA preparation precisely from H-5 minutes to H-1 minute. If previously planned, this can be executed with degraded communications.

(2) An event-driven action might be inserting Company B into the alternate LZ if Company A (the lead company) makes enemy contact on the primary LZ. If previously planned, this event will occur properly without the need for lengthy radio communications by the AATFC.

c. **Decentralize control.** Although it is centrally planned, air assault execution is decentralized. Subordinate commanders should be given the maximum possible freedom of action (consistent with safety and mission accomplishment considerations) to ensure mission accomplishment.

d. **Establish air assault radio nets.** Radio nets to facilitate ground-to-ground, air-to-air, and ground-to-air communications are established to provide for the timely flow of information and redundancy in capability.

2-16. Roles of key personnel and critical modes.

Key players and communications modes in air assault C^2 are:

a. **Air assault task force commander.** The AATFC is normally an infantry brigade or battalion commander who is the overall AATF commander. His presence and role ensures a unity of command throughout the operation. As in any operation, he must move where he can see the battlefield and control the operation. In situations where the enemy allows, he would be airborne during the movement and insertion phases. At other times, he fights the battle from a tactical command post deployed well forward.

b. **Air assault task force S3.** The AATF S3 assists the AATFC in C^2. He normally mans the AATF tactical command post (CP) when the AATFC is airborne.

c. **Air mission commander.** The AMC is an aviation unit commander or his designated representative. He is responsible for receiving and executing the AATFC's guidance and directives and for controlling all aviation elements for the AATFC. His presence ensures unity of effort for all supporting aviation assets. The AMC employs attack helicopters and artillery along the flight route and "fights the battle" from PZ to LZ while keeping the AATFC informed.

d. **Aviation liaison officer.** Although the LO's most critical role is fulfilled during the planning phase, he can be a valuable team member in C^2 if he has access to adequate radio equipment. When he is radio equipped, the AATFC may employ the aviation LO at a critical point to assist in coordinating the execution of the operation.

e. **Lift flight lead.** He leads the lift aircraft along the route(s) of flight, adjusting airspeed as necessary to meet preplanned artillery SEAD and preparatory fire schedules.

f **Air reconnaissance and attack helicopter battle team captains.** These are air reconnaissance or attack helicopter platoon leaders or troop (company) commanders who are responsible for the C^2 of their respective elements. They normally respond to the AMC during the movement phase and to the AATFC as subordinate maneuver unit commanders after completion of the air assault insertion.

g. **Pickup zone control officer.** A pickup zone control officer (PZCO) is designated for each pickup zone to be used. He organizes, controls, and coordinates operations in the PZ and "pushes" elements out. He operates on the combat aviation net (CAN) and is prepared to assist in executing needed changes. He is the key individual during night operations or when multiple subordinate elements are being lifted from the same PZ.

h. **Subordinate unit commanders.** Subordinate unit commanders normally function as they would in any other infantry task force. Each must be prepared, however, to receive other elements for movement.

i. **Tactical command post.** The tactical command post (TAC CP) provides C^2 for the execution of air assault operations. It must be mobile and well forward. It is normally air assaulted into the objective area soon after the initial echelon, the enemy situation permitting. A C^2 helicopter may serve as a TAC CP if enemy air defense systems allow.

j. **Main command post or tactical operations** center. The main CP or tactical operations center (TO C) provides control of combat operations when the TAC CP is not deployed, and provides planning for future operations and coordination for support. Functions of the main CP are:

(1) Monitors current operations and maintains current enemy and friendly situations.

(2) Gathers and disseminates intelligence.

(3) Keeps higher and adjacent organizations informed of the friendly situation; submits

recurring reports.

(4) Provides liaison to higher and adjacent organizations.

(5) Coordinates combat support, close air support (CAS), aviation (AVN), engineer (EN), ADA, and advises the commander on the use of combat support for current and future operations.

(6) Monitors airspace management.

(7) Continues planning for future operations; oversees the preparation of all contingency plans.

(8) Issues combat orders and warning orders as necessary.

k. **Rear command post.** The rear CP is normally located in the field trains, and coordinates all logistical and personnel operations and requirements. The administration logistics center is the nerve center of the rear CP and coordinates CSS for the AATF.

2-7. Communications.

a. Command and control within the AATF are executed with a variety of communications means to span the full spectrum of air assault operations. To support an AATF over a widely dispersed area, emphasis is placed on compact, lightweight, air transportable, and long-range equipment, A heavy reliance is placed on single channel communications such as very high frequency (VHF)/frequency modulation (FM), high frequency (HF)/single side band (SSB), and tactical satellite communications (TACSATCOM).

b. Real time C^2 capabilities will be constrained by the availability of portable, reliable, and secure communications. An AATF must depend largely upon a single channel radio because of its flexibility, range, and speed of set up.

c. Subordinate elements in the AATF may range beyond multichannel capabilities and radio transmissions, and transmissions may be unintelligible due to enemy electronic countermeasures (ECM). As a result, subordinate commanders of the AATF will be required to make decisions sometimes without being in contact with the AATFC.

d. As the AATF fights the battle and distances become extended, communications for C^2 become less sophisticated. The AATF must make extensive use of airborne or unattended FM retransmission, amplitude modulation (AM) capabilities, and TACSTATCOM. Ground or air messengers should be used when possible.

2-18. Radio nets.

A dynamic mix of air-to-air, air-to-ground, and ground-to-ground radio nets is used to provide the necessary responsiveness and flexibility for air assault C^2. Radio nets commonly employed during air assault operations are:

a. Air assault task force command net. This is an FM command net (ground-to-ground) for an operation. It is normally secure and used by the AATFC to communicate with his subordinate maneuver commanders.

b. Combat aviation net. This is an FM radio net dedicated to air-to-ground coordination during air assault operations. All aviation elements monitor this net as do the remainder of the AATF elements before and during air movements. Although the CAN may serve as an alternate task force (TF) command net, it must be dedicated primarily to communications between aircraft and the lifted unit. Its use for that purpose ensures that mission and situation changes can be quickly passed to supporting aircraft and that the AATF command net remains clear for use by the AATFC and his subordinate commanders.

c. Air battle net (ABN). This is an ultra high frequency (UHF) air-to-air command net dedicated to communications between the AMC and all aviation element leaders. All aviation elements monitor this net and receive instructions from the AMC or the AATFC when he is airborne. This net is normally operated on the lift unit's UHF command frequency if a dedicated ABN is not listed in the communications-electronics operation instructions (CEOI).

d. Fire **support** net. This is an FM net operated by the AATF fire support coordinator (FSCOORD). All aviation elements must have access to this net to facilitate calls for fire during movements, insertions, and extractions. An artillery quick-fire net would normally be used when a supporting battery is dedicated to an operation.

e. Aviation internal net. These are VHF nets operated by each aviation element leader for his own internal use. Use of VHF radios provide each element leader with a dedicated frequency with which to direct and control individual aircraft, teams, or platoons, and to communicate with air traffic control (ATC) authorities. Figure 2-2 depicts the inter-relationships of these radio nets.

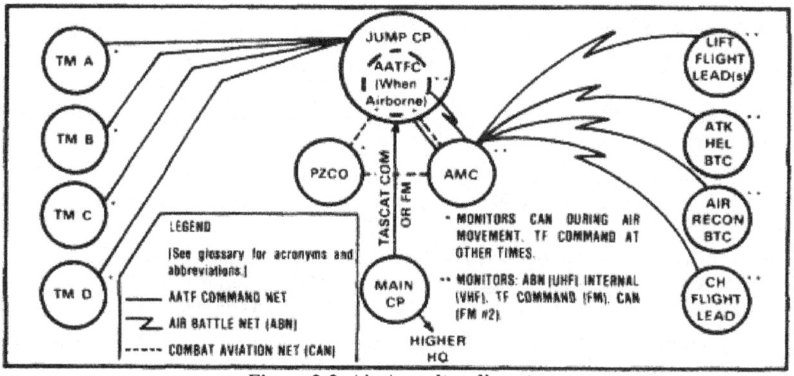

Figure 2-2. Air Assault radio nets.

CHAPTER 3

Operations Planning

Section I

ESTIMATE PROCESS

3-1. General.

a. Successful air assault execution is based on a careful analysis of METT-T and detailed, precise reverse planning. Five basic plans that comprise the reverse planning sequence are developed for each air assault operation. They are:

- The ground tactical plan.
- The landing plan.
- The air movement plan.
- The loading plan.
- The staging plan.

These plans should not be developed independently. They are coordinated and developed concurrently by the AATF staff to make best use of available time. The ground tactical plan is normally developed first and is the basis from which the other plans are derived.

b. Planning for air assault operations requires time - time to plan, time to prepare, and time to brief. The AATF uses the sequence of command and staff actions and troop leading procedures common to other combat operations.

c. Planning for air assault operations is as detailed as time permits and should include completion of written orders and plans. Within time constraints, the AATFC must carefully evaluate capabilities and limitations of the total force and develop a plan that ensures a high probability of success.

d. Often, however, the fleeting nature of tactical opportunities does not permit adequate planning time and the development of detailed written plans and orders. If time is limited, planning steps may be compressed or conducted concurrently; detailed written plans and orders may be supplanted by standing operating procedure or lessons learned in previous training. Previous training and the development of SOPs cannot be overemphasized. Units cannot expect to successfully conduct air assault operations, particularly with compressed planning time, without the benefit of previous training.

e. Many routine tasks related to air assault operations are accomplished above the AATF level. The division is the lowest echelon that can allocate assets, assign appropriate missions, gather required data, and analyze capabilities. For this reason, when an air assault mission is assigned by division, or higher level command, that headquarters begins the planning process. The division uses its resources to gather data and provides planning information to lower echelons, or division may complete the planning tasks itself. When the division does these tasks, subordinate commanders can expend their limited time to accomplish other key planning tasks.

f. When an infantry unit is given an air assault mission, the assigning echelon provides the latest extended weather forecast, up-to-date intelligence (with emphasis on known or suspected enemy air defense systems), initial fire planning, and many of the terrain considerations relevant to the operation. Additional information that is not provided may be requested and/or completed by the AATF. All echelons attempt to reduce the planning burden of subordinate units.

g. The battalion is the lowest level that has sufficient personnel to plan, coordinate, and control an air assault operation. When company-size operations are conducted, the bulk of the planning takes place at battalion and higher headquarters.

h. All tactical estimates used in troop leading procedures employ the factors of METT-T. The METT-T provides data that is analyzed using the estimate process and from which a decision is made. Applying the factors helps the commander isolate and address significant considerations that affect the mission. The factors of METT-T are considered in each phase of the estimate.

3-2. Mission.

Mission analysis is conducted early-on in the estimate process. The mission involves the critical tasks that must be performed. The tasks are either specified tasks stated by the order or implied tasks that the commander must deduce. Mission analysis determines not only what must be accomplished, the intent of the commander ordering the mission (the why of the operation), and the limitations (when, where, how) placed by the higher headquarters, but is the basis for deciding on task organization. Once the mission is analyzed and deductions are made, all other factors are considered in terms of their impact on the mission. It is therefore imperative that the mission be understood before continuing the estimate.

3-3. Enemy.

The examination of enemy factors should be as detailed as possible depending on the time available.

a. General factors to consider are:

(1) Identification - who is he? Size and type of unit.

(2) Location - where is he and where is he going?

(3) Disposition - how is he organized; what are his formations?

(4) Strength - his versus friendly forces.

(5) Morale - esprit, experience, state of training, regular or reserve.

(6) Capabilities - electronic warfare, NBC, air defense, airborne, airmobile, attack helicopters, mobility (in comparison to the air assault unit).

(7) Composition - armor, infantry (motorized or light), artillery, combat support.

(8) Probable courses of action - what is his likely mission or objective, and how will he probably achieve it?

b. When planning an air assault operation, the following factors about the enemy must be considered:

(1) His air defense weapons and capability.

(2) His mobility; particularly his ability to react to an air assault insertion.

(3) His NBC capability; particularly his ability to influence potential flight routes and landing zones.

(4) His capability to interdict or interrupt air assault operations with his helicopters or fixed-wing aircraft.

(5) His EW capability.

3-4. Terrain.

In all military operations, terrain analysis is conducted by the criteria described in the term OCOKA:

- Observation and fields of fire.
- Cover and concealment.
- Obstacles and movement.
- Key terrain.
- Avenues of approach.

In air assault operations, these factors must be analyzed in terms of their effect on the air assault force during pick up, air movement, insertion, and movement to the final objective, and in terms of OCOKA's overall influence on aviation operations.

a. **Observation and fields of fire.** These considerations relate to both enemy and friendly forces and, for air assault operations, include:

(1) Enemy visual observation and/or electronic surveillance of PZS, flight routes, and LZs.

(2) Enhanced friendly observation provided by scout and aerial field artillery observation helicopters.

(3) Ease of navigation along flight routes particularly for night or adverse weather operations.

b. **Cover and concealment.**

(1) Terrain masking for nap-of-the-earth (NOE) flight routes and insertions.

(2) Covered firing positions for attack helicopters.

(3) Landing zones which offer infantry cover and concealment following insertion.

c. **Obstacles and movement.** While most obstacles can be bypassed by air assault forces, obstacles which affect the ground tactical plan must be considered.

d. **Key terrain.** Key terrain is mission-dependent; however, in air assault operations key terrain is not limited to that which influences the ground tactical plan. It must also be analyzed in terms of.

(1) Pickup zones and/or landing zones.

(2) Flight routes.

(3) Attack helicopter battle positions.

(4) Occupation by enemy ADA assets.

(5) Potential forward area rearming and refueling points (FARP).

e. **Avenues of approach.** Air and ground avenues of approach are considered in both offensive and defensive operations from friendly and enemy viewpoints. A good avenue of approach for air assault forces offers:

(1) A reasonable degree of mobility and few if any natural obstacles to the aircraft.

(2) Little or no canalization.

(3) Terrain masking that decreases effectiveness of enemy air defense weapons.

(4) Cover.

(5) Concealment.

(6) Good lines of communication and logistics.

(7) Ease of linkup with other forces when appropriate.

f. **Weather and visibility.** Weather information is analyzed for trends (Figure 3-1). If the operation begins in marginal weather, the commander must consider the possibility that it will deteriorate below acceptable limits during the operation. This may result in an interruption of helicopter support and require changes in planned operations. Considerations include:

(1) Fog, low clouds, heavy rain, and other factors that limit visibility for aviators.

(2) Illumination and moon angle during night vision goggle (NVG) operations.

(3) Ice, sleet, and freezing rain that degrades aerodynamic efficiency.

(4) High temperatures and/or density altitudes that degrade aircraft engine performance and lift capability.

(5) Darkness, normally an advantage to well-trained aviators and soldiers.

(6) High winds (large gust spreads)

(7) Weather conditions that create hazards on PZs and LZs, such as blowing dust, sand, or snow.

FORMAT	Maximum winds
Weather forecasts are received in the following format:	Observation helicopter (OH) 30 knots.
Ceiling.	Utility helicopter (UH) 40 knots.
Visibility.	Cargo helicopter (CH) 60 knots.
Weather (e.g., clear, fog, rain, snow).	**NOTE: Gusting winds, in excess of 15 knots over the lull wind, may preclude UH usage. Significant weather patterns (which limit operations) are moderate turbulance and icing.**
Additional information as requested by the S2.	
CONSIDERATIONS	**Extremes Limiting Tactical Air (TACAIR)**
Allowable Weather Limits	Ceiling 1,000 feet.*
(Applicable to combat operations and tactical training at a military airfield.)	Visibility 2 miles.*
Visibility 1/2 mile.	***NOTE: Operational design of a A-10, close air support (CAS) aircraft. Other type aircraft require better weather conditions.**
Ceilings Clear of clouds.	

Figure 3-1. Weather data.

3-5. Troops available.

a. The AATF should have enough combat power to seize initial objectives and protect the LZs until follow-on echelons arrive in the objective area.

b. Assault (lift) helicopter capability is the single most important variable in determining how much combat power can be introduced into the objective area.

c. Aircrew endurance must be considered. For planning purposes, the AATFC should consider eight hours a day and four hours of night flying to be a safe limit for aircrews. If those limits are exceeded during a single period, then degraded aircrew performance can be expected on the following days.

3-6. Time available.

The following items are critical to the operations:

a. The time available for preparation, planning, and rehearsals is crucial. Air assault planning must be centralized and precise, and normally takes more time than that for other operations.

b. Normally, additional planning time must be allotted for night operations and those involving multiple PZs and/or multiple LZs.

c. The AATFC must allow adequate time to ensure that all subordinates, particularly aircrews, are thoroughly briefed. Briefing time is significantly reduced by viable SOPs and previous training.

d. The AMC must be provided time to brief and totally integrate all aviation units.

Section II

SECURITY AND CONTROL

3-7. General.

Operational security (OPSEC) is the protection of military operations and activities from enemy exploitation. It includes those actions taken to deny the enemy information about planned, ongoing, and completed operations. Effective OPSEC helps maintain surprise in air assault operations and is a tactical imperative because of the density of helicopters involved, the reliance on radio communications, and the potential for catastrophic losses if plans and operations are compromised. An air assault task force makes a lucrative target for enemy air defense, air, and artillery systems. Every reasonable effort must be made to avoid disclosing intended locations and time(s) of air assault operations and thus losing the element of surprise. A "telegraphed punch" can be catastrophic to air assault operations.

3-8. Countermeasures.

The AATF is vulnerable to enemy intelligence gathering methods and must counter the Threat with OPSEC measures including:

a. **Signal security.** This includes:

(1) Radio listening silence when possible.

(2) Use of low power transmission,

(3) Use of directional antennas.

(4) Proper use of brevity codes and radio procedures.

(5) Secure communications equipment.

b. **Information security.** Plans and orders must be safeguarded; information must be limited to those with a need to know.

c. **Deception operations.** It is often appropriate and necessary for the AATFC to employ deception operations to ensure the success of his mission. They may include:

(1) Firing false artillery preparation.

(2) Making false insertions.

(3) Maneuvering forces to other areas away from the objective.

(4) All other infantry deception techniques.

d. **Passive security.** This includes camouflage, countersurveillance, noise and light discipline, warning devices, and rapid troop insertions, plus:

(1) Tactical dispersion of helicopters and units; PZs and LZs must not be congested.

(2) Keeping helicopter FARPs well to the rear and displacing them frequently.

(3) Marking PZs and LZs to avoid compromise.

(4) Careful planning of flight routes and altitudes, using terrain masking to deny the enemy direct observation.

(5) Using multiple PZs, LZs, and routes.

(6) Using the speed and maneuverability of helicopters to gain surprise.

e. **Active security.** These measures include employment of patrols, observation post, and reconnaissance. In air assault operations, active security measures include:

(1) Employment of air reconnaissance units to provide early warning around LZs and objective areas.

(2) Suppression of enemy intelligence gathering capability.

(3) Air force reconnaissance capability.

(4) Employment of long-range surveillance units (LRSU) of the division's military intelligence battalion.

3-9. Army airspace command and control.

Army airspace command and control (A2C2) are considered early in the planning stage to resolve conflicts and provide for the safe movement of friendly aircraft. Controlling conflict of airspace use should be executed by SOP, directives, and other passive measures because radio communications and positive control of all aircraft may not be possible during actual operations. (See Appendix G for a discussion of A2C2).

Section III

GROUND TACTICAL PLAN

3-10. General.

The foundation of a successful air assault operation is the commander's ground tactical plan, around which subsequent planning is based. The ground tactical plan specifies actions in the objective area to ultimately accomplish the mission and address subsequent operations.

3-11. Elements of the ground tactical plan.

a. The ground tactical plan for an air assault operation contains essentially the same elements as any other infantry attack but differs in that it is prepared to capitalize on speed and mobility in order to achieve surprise. Assault echelons are placed on or near the objective and organized so as to be capable of immediate seizure of objectives and rapid consolidation for subsequent operations. If adequate combat power cannot be introduced quickly into the objective area, then the air assault force must land away from the objective and build up combat power. The air assault force then assaults like any other infantry unit and the effectiveness of the air assault operation is diminished.

b. The scheme of maneuver may assume a variety of possibilities depending on the commander's evaluation of METT-T including, in particular, the availability of LZs in the area. The plan should include:

(1) Missions of all task force elements and methods for employment.

(2) Zones of attack, sectors, or areas of operations with graphic control measures.

(3) Task organization to include command relationships.

(4) Location and size of reserves.

(5) Fire support to include graphic control measures.

(6) Combat service support.

NOTE: This plan is prepared by thE AATF staff with input from all task force elements and is in sufficiEnt detail to facilitate understanding by subordinate commanders. It is imperative that all aircrews know this ground tactical plan and the ground commander's intent.

Section IV

THE LANDING PLAN

3-12. General.

a. The landing plan must support the ground tactical plan. This plan sequences elements into the area of operations, ensuring that units arrive at designated locations and times prepared to execute the ground tactical plan.

b. General considerations in developing the landing plan are:

(1) The availability, location, and size of potential LZs are overriding factors.

(2) The AATF is most vulnerable during landing.

(3) Elements must land with tactical integrity.

(4) Troops are easily disoriented if the briefed landing direction changes, and they are not kept informed.

(5) Initially, there may be no other friendly units in the area. The AATF must land prepared to fight in any direction.

(6) The landing plan should offer flexibility so that a variety of options are available in developing a scheme of maneuver.

(7) Supporting fires (artillery, naval gunfire, CAS, attack helicopters) must be planned in and around each LZ.

(8) Although the objective may be beyond the range of supporting artillery fire, artillery or

mortars may be brought into the landing zone(s) early to provide fire support for subsequent lifts and on the objective.

(9) The plan should include provisions for resupply and medical evacuation by air.

3-13. Selection of landing zones.

a. **Selection criteria.** Landing zones are selected by the AATFC or his S3 with technical advice from the AMC or his liaison officer. They do so using the following criteria:

(1) **Location.** It can be located on, near, or away from the objective, depending on the factors of METT-T.

(2) **Capacity.** The size determines how much combat power can be landed at one time. This also determines the need for additional LZs or separation between serials.

(3) **Alternates.** An alternate LZ should be planned for each primary LZ selected to ensure flexibility.

(4) **Enemy disposition and capabilities.** Enemy troop concentrations, air defenses, and their capability to react to an AATF landing nearby are considered when selecting an LZ.

(5) **Cover and concealment.** Landing zones are selected that deny enemy observation and acquisition of friendly ground and air elements while they are en route to and/or from (and in) the LZ.

(6) **Obstacles.** If possible, the AATF should land on the enemy side of obstacles when attacking and use obstacles to protect LZs from the enemy at other times. Landing zones must be free of obstacles. Engineers must be organized for contingency breaching of obstacles.

(7) **Identification from the air.** Landing zones should be readily identifiable from the air. They should be marked with chemical lights, preferably infrared type, if the assault is conducted with personnel wearing night vision goggles.

NOTE: This assumes the presence of a friendly reconnaissance unit that has reconned and marked the LZs.

(8) **Approach and departure routes.** Approach and departure flight routes should avoid continued flank exposure of aircraft to the enemy.

(9) **Weather.** Reduced visibility or strong winds may preclude or limit the use of marginal LZs.

b. **Options to consider.** If there are options available in selecting LZS, the ones that best facilitate mission accomplishment are chosen. This choice involves whether to land on or near the objective, or to land away from it and maneuver forces on the ground to the objective. Factors considered in making that determination are:

(1) **Combat power.** This includes maneuver elements, firepower, and combat support assets that can be introduced into the area early in the operation (usually dependent upon the number of aircraft employed and availability of suitable LZs).

(2) **Enemy.** This includes enemy strength and disposition in and around the objective area, to include air defense systems.

(3) **Surprise.** This is a goal that may be attained by careful use of terrain, cover and concealment, darkness, or reduced visibility created by weather or smoke. Surprise is sometimes achieved by landing on the objective.

(4) **Time.** Time that is available for mission accomplishment. Limited time to complete the mission generally favors landing on or near the objective.

c. **Landing zone characteristics.** In addition to deciding where to land in relation to the objective, a decision is made on whether to use a single LZ or multiple LZs.

(1) **Advantages of a single LZ.**

(a) Allows concentration of combat power in one location (if the LZ is large enough).

(b) Facilitates control of the operation.

(c) Concentrates supporting fires in and around the LZ. Firepower is diffused if more than one LZ preparation is required.

(d) Provides better security for subsequent lifts.

(e) Requires fewer attack helicopters for security.

(f) Reduces the number of flight routes in the objective area, making it more difficult for enemy intelligence sources to detect the air assault operation.

(g) Centralizes any required resupply operations.

(h) Concentrates efforts of limited LZ control personnel and engineers on one LZ.

(i) Requires less planning and rehearsal time.

(2) **Advantages of multiple LZs.**

(a) Avoids grouping assets in one location and creating a lucrative target for enemy mortars, artillery, and CAS.

(b) Allows rapid dispersal of ground elements to accomplish tasks in separate areas.

(c) Reduces the enemy's ability to detect and react to the initial lift.

(d) Forces the enemy to fight in more than one direction.

(e) Reduces the possibility of troop congestion in one LZ.

(f) Eliminates aircraft congestion on one LZ.

(g) Makes it difficult for the enemy to determine the size of the air assault force and the exact location of supporting weapons.

NOTE: If the objective is designated by a number, the LZ should be designated by a letter or code word to avoid confusion and preclude mix-ups. This avoids having an objective (OBJ) and LZ with the same designator; for example, LZ 1 and OBJ 1.

3-14. Landing formations.

Aircraft formations on the LZ should facilitate off-loading and deployment for the assault. The number and type of aircraft, and the configuration and size of the LZ, may dictate the formation. Because contact is expected in the LZ, elements are landed ready to employ fire and movement. An LZ formation should not be a picture-perfect formation with standardized distances between aircraft. Landing aircraft rapidly select a SAFE landing area as close to concealment as possible to reduce troop exposures. (NOTE: If possible, the PZ formation is the same.) This provides troops a preview of the LZ and gives them an idea of where they will be located (upon landing) in relation to other elements. (For further discussion of air assault formations and drills, see Appendix C.)

3-15. Fires to support the landing plan.

a. Frequently, it is desirable to make the initial assault without preparatory fires in order to achieve tactical surprise. However, preparations are planned for each LZ so that they can be fired if needed.

b. Planned fires for air assault operations should be intense and short but with a high volume of fire to maximize surprise and shock effect. The fires should end just before the first assault element's landing.

c. When developing fire support plans, consideration is given to:

(1) **Deception.** False preparations are fired into areas other than the objective or LZ area.

(2) **Loss of surprise.** A preparation of long duration may reduce the possibility of surprise.

(3) **Availability of fire support.** The fire support officer (FSO) considers assets that can fire a preparation and coordinates with the artillery unit to arrange the preparation. Preparations by tactical aircraft or attack helicopters may be the only viable alternative in many cases.

(4) **Significant targets.** A known or suspected enemy force, regardless of size, warrants an LZ preparation.

(5) **Obstacles to landing and maneuver.** Some ordnance used in preparation (artillery, bombs, napalm) can cause craters, tree blowdown, fires, and LZ obscuration and therefore may not be desirable.

(6) **Scheduling fires.** Fires are scheduled to be lifted or shifted to coincide with the arrival times of aircraft formations.

(7) **Positive control measures.** Control measures must be established for lifting or shifting fires.

Section V

AIR MOVEMENT PLAN

3-16. General.

a. The air movement plan is based on the ground tactical and landing plans. It specifies the schedule and provides instructions for air movement of troops, equipment, and supplies from PZs to LZs. It also provides coordinating instructions regarding air routes, air control points, and aircraft speeds, altitudes, and formations. The planned use of attack helicopters, to include security and linkup locations (if different from PZ), should also be included in the air movement plan. (When operations involve multiple lifts from the same PZ, a lift table is prepared to ensure lifts are properly organized.)

b. The air movement plan is normally developed in coordination with the AMC, or the aviation liaison officer, who provides technical assistance and recommendations.

3-17. Development of tentative flight routes.

Tentative flight routes are developed to control, protect, and sequence aircraft movement. Careful consideration is given to the terrain and enemy forces. The AATF S2, S3, and AMC assist the AATFC in developing flight routes. The basic methods of developing tentative flight routes is by map study or by photo review when time permits, considering the locations of friendly units, enemy dispositions and air defense systems, and PZs and/or LZs. A flight route consists of a start point (SP), release point (RP), and a flight path between the two. The fire support plan should include fire planning along the flight routes.

3-l8. Designation of start point and release point.

The first step is to identify tentative SPs and RPs (Figure 3-2). The distance from the PZ to the SP should be no less than three to five kilometers to allow aircraft to achieve the desired airspeed, altitude, and formation after lift-off. The distance from the RP to the LZ should allow the flight leader to reconfigure the formation and execute a tactical formation landing. In locating SPs and RPS, the following considerations apply:

a. Locate three to five kilometers from PZs and LZs respectively. This allows two to three minutes flying time for coordination of the flight's en route procedures.

b. Locate according to weather, obstacles, and enemy positions.

c. Locate to facilitate lift-off and landing into the wind by the best flight path.

3-19. Air movement table.

a. The air movement table:

(1) Contains aircraft allocations.

(2) Designates number and type of aircraft in each serial.

(3) Specifies departure point, route to and from loading area, and loading, lift-off, and landing times.

b. The table is prepared jointly by the AATF staff and aviation personnel and it is completed in detail since it serves as the primary movement document.

NOTE: The table controls AATF movement from PZ to LZ as air assault forces fly to the LZ utilizing radio listening silence, if possible.

c. The table ensures that all personnel, equipment, and supplies are accounted for in the movement, that each aircraft is fully loaded, correctly positioned in the flight, and directed to the right LZ (Figure 3-3). The table must also include the refuel schedule for all lifts, if required.

In this situation, the SP is within 3 to 5 kilometers of the PZ. It provides sufficient air maneuver space for helicopters to arrive at the SP in the prescribed en route formation at the proper airspeed.

The SP is located at a recognizable topographic feature for ease of identification. It facilitates lift-off from the PZ and heading into the wind. It is located to avoid known flight obstacles and enemy locations.

The RP is also 3 to 5 kilometers from the designated LZ. It is used as the final checkpoint for coordination of landing instructions and lifting or shifting of preparatory fires.

The distance from RP and the LZ is used by helicopter pilots to shift into landing formation, reduce speed, and land on their assigned LZ. The RP is also located at a recognizable topographic feature and avoids obstacles and enemy locations.

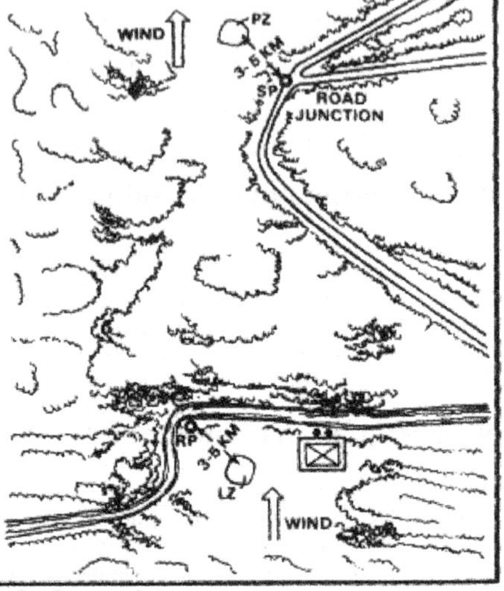

Figure 3-2. Start and release points.

EXPEDIENT MOVEMENT TABLE

COMPANY C

PZ PRIMARY - BLUE ALTERNATE - BLUE 1 ____ Designated PZ(s) and LZ(s).
LZ PRIMARY - 1 ALTERNATE - 2

AIRCRAFT - 18 1 UH-1 - 121 COMBAT SUPPORT AVIATION COMPANY ____ Number and type of helicopters available to the particular elements (to include unit and/or call signs).

ARRIVE IN PZ H-21 ____ Time of arrival at PZ(s).

LIFT-OFF FROM PZ H-18 ____ Time of lift-off at PZ(s).

LAND IN LZ H-HOUR (1000 HRS) ____ Landing time at LZ(s).

FLIGHT ROUTE PRIMARY - APPLE, ALTERNATE - ORANGE ____ Designated flight route(s).

ACL - 8 ____ ACLs (correct for type aircraft, according to density altitude at lift-off).

Figure 3-3. Expediant movement table.

3-20. Development of flight routes.

a. Flight routes are developed based on tactical and technical factors. It may be necessary for a route to pass through an adjacent unit's sector. When that is the case, approval from that unit is obtained and coordination is made. Regardless of route direction or location, certain criteria are considered.

b. Seldom are all characteristics present in any one situation; one or more may have to be omitted. Flight routes:

(1) Are as short as possible, consistent with other considerations.

(2) Avoid turns in excess of 45 degrees, when formation flying is required, to facilitate control of the aircraft formation.

(3) Provide terrain masking to deny exposure to enemy observation, direct fire weapons, and radar acquisition, if possible.

(4) Provide cover when terrain permits, placing terrain mass and/or vegetation between the enemy and the aircraft.

(5) Provide for ease of navigation (day or night).

(6) Avoid masking friendly fires, particularly supporting artillery.

(7) Avoid known enemy units and air defense positions.

(8) Avoid overflight of built-up areas.

3-21. Flight corridor.

a. When there is competition for airspace, it may be necessary to modify the flight route(s) and designate a flight corridors). The corridor reserves airspace around a flight route for AATF use, and prevents artillery, tactical air (TACAIR), and other elements from firing or flying through when it is in use.

b. Authority to establish a flight corridor is obtained from the brigade and/or division commanders). Designated flight corridors are coordinated through airspace management channels. This ensures that airspace within corridors is not violated.

c. The corridor begins as a flight route and is then modified as required. The size of corridors varies. Normally, they extend 200 to 300 meters on either side of the designated flight route, and 500 feet above and below the route flight altitude.

d. Helicopter formations operating at terrain flight (low) altitudes do not require minimum altitude corridor designations. The upper air limit of the corridor may vary and would be specified by the headquarters establishing it.

e. If it is necessary to restrict the operational area to only those aircraft directly involved in the air assault operation, a restricted area can be established by the airspace management element.

3-22. Flight axis.

a. The flight axis is another variation of the flight route. It is a flight route that has width (like the corridor) but does not have airspace reserved to a specific altitude (as does the corridor).

b. The flight axis permits deviation laterally along the flight route but does not restrict the employment of other assets. It gives the AMC a choice in selecting en route formations and freedom to alter direction without coordinating a new flight route.

3-23. Expedient flight routes.

These routes are established with checkpoints. If time is not available to develop and disseminate PZs, LZs, and flight route information, the commander can define an expedient route by reference to checkpoints (Figure 3-4).

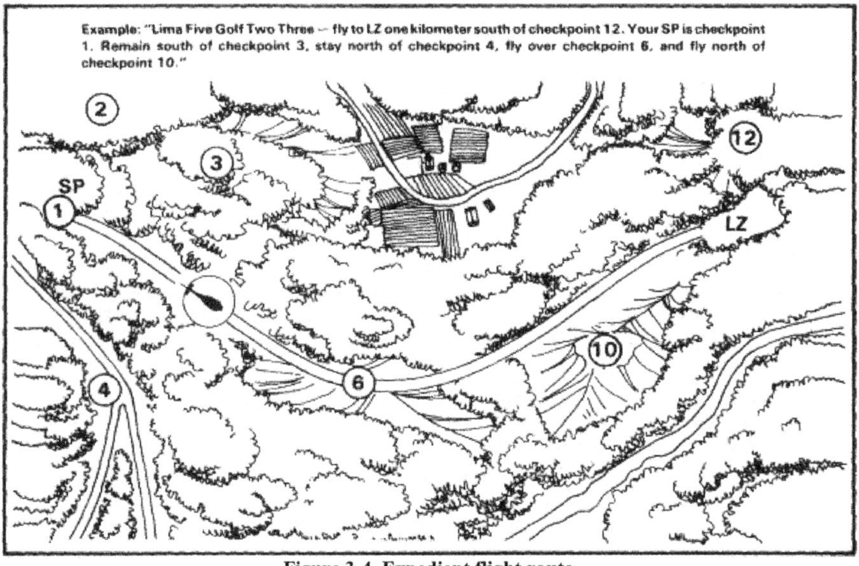

Figure 3-4. Expedient flight route.

3-24. Flight route control measures.

a. Control measures assist in navigation and provide control to ensure the AATF arrives in the LZ on time and in sequence.

b. Air control points (ACP) designate each point where the flight route changes direction (Figure 3-5). They include readily identifiable topographic features or points marked by electronic navigational aids.

A route may have as many ACPs as necessary to control the air movement. The SPs and RPs are also air control points.

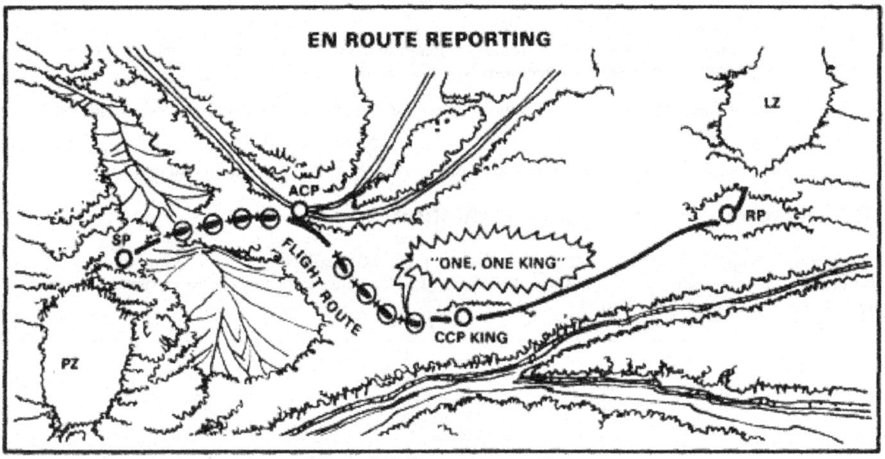

Figure 3-5. En route reporting.

c. An ACP may be further designated as a communication checkpoint (CCP). A CCP is a point along the flight route that serial commanders report to the AMC. Radio transmissions are made only when necessary. If a report is required, the transmission is short. This is possible by using codes. For example, the short radio transmission, "One, One King," could mean that the first serial of lift I is crossing CCP King.

3-25. Designation of routes.

a. Once tentative flight routes are identified, they are designated for use by each unit. When large groups of aircraft are employed, dispersion is achieved by using multiple routes. However, with large serials it is often necessary to use fewer routes, or even a single route, in order to concentrate available supporting fires. Also, the number of alternate and return routes may be limited.

b. Primary, alternate, and return routes to be used by each subordinate unit are designated. When selecting routes, the following factors are considered:

(1) **Interference with ground action.** Overflying ground elements may interfere with their supporting fire. Flight routes should be clear of the gun-target line when possible.

(2) **Support of landing plan.** To reduce vulnerability of the air assault force, flight routes should facilitate rapid approach, landing, and departure from selected LZs.

(3) **Enemy ground and air capabilities.** Selected flight routes make maximum use of terrain, cover, and concealment to minimize enemy observation and target acquisition.

(4) **Available fire support.** Flight routes allow support from all available resources.

(5) **Available air cover.** Flight routes are identified in order to provide air cover for friendly forces en route.

(6) **Weather conditions.** Flight routes remain usable based on prevailing weather during execution of the air assault operation.

(7) **Terrain.** Flight routes use terrain to maximum advantage to reduce vulnerability of the aircraft formations.

(8) **Time (distance) from** PZ **to LZ.** Flight routes are as short as possible to reduce flying time.

3-26. Dissemination of route information.

Maps or overlays containing flight route information are prepared at AATF headquarters and disseminated to subordinate and support units. (Overlays are often used.) Flight routes and corridors are designated by a letter, number, or word (Figure 3-6).

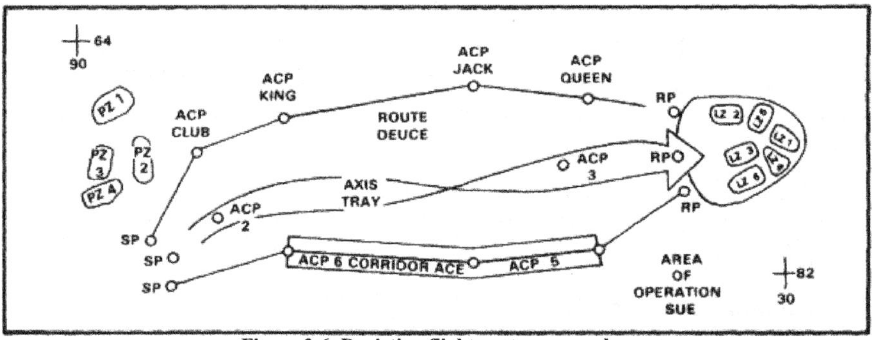

Figure 3-6. Depicting flight routes on overlay.

3-27. En route formations.

The flight's formation is dictated by the terrain, enemy situation, and the degree of control required. Regardless of the specific formation, aircraft are staggered and the distance between them varies according to the terrain being crossed. The AMC and/or flight leader selects the en route formation. The aircraft land in the formation specified by the air movement table (Figure 3-7).

ANNEX B (AIR MOVEMENT TABLE) to OPORD 8, TF 1-67 (Classification) Copy No ______ of ______ Copies
Time Zone Used Throughout the Order: ALPHA
Hqs, 1 Bn, 67th Inf
101500 Jan 8 __

P - Primary
A - Alternate

AVN UNIT	LIFTED UNIT	LIFT	SERIAL	LOADS	PICKUP ZONE	LOADING TIME	LIFT-OFF TIME	SP TIME	RP TIME	LANDING ZONE	LANDING TIME	REMARKS
	A	1	1	1-4	ALPHA S/L ALTN BRAVO		H-12	H-10	H-2	GREEN S/L ALTN YELLOW	H-HOUR	BUMP 4, 7, 10, 16 FLT RT P-EAGLE A-HAWK
			2	5-8								
			3	9-12	BRAVO S/L					GREEN S/L		
			4	13-16								
	B	2	1	1-4	CHARLIE S/K					GREEN S/K		BUMP 4-8, 12, 16 FLT RT P-HAWK A-BAT
			2	5-8								
			3	9-12								
			4	13-16								

[1] NOW REFER TO CODE-NUMBERED AIR MOVEMENT TABLE (PAGE 3-37) TO SEE HOW THIS INFORMATION APPEARS IN ITS ENTIRETY

Figure 3-7. Flight routes and lifted units.

3-28. Terrain flight modes.

A specific en route flight altitude is not designated. Pilots may use one of three terrain-flight modes as dictated by the mission and the Threat (Figure 3-8).

a. **Nap-of-the-earth flight.** This is flown at varying airspeeds and altitudes as close to the earth's surface as possible while following the contours of the earth. It is a weaving flight path that remains oriented along the general axis of movement and takes advantage of terrain masking.

b. **Contour flight.** This is flown at low altitude conforming generally to the contours of the terrain. The flight is characterized by varying altitudes and varying airspeed.

c. **Low level flight.** This is flown at low altitude, with constant heading, airspeed, and altitude to facilitate speed and ease of movement while minimizing detection. This mode of flight is normally used only in rear areas.

d. **Factors affecting flight attitude.** These include:

(1) **Enemy.** The greater the enemy air defense threat, the lower the flight altitude.

(2) **Terrain.** Aircraft must clear all terrain obstacles and still reduce exposure to enemy air defense weapons and observation.

(3) **Navigation.** It is usually easier to navigate at higher altitudes, but the risk of detection by the enemy is greater.

(4) **Weather.** Ground fog, or haze, requires higher altitudes whereas a low ceiling requires lower altitudes.

(5) **Flight distance.** If the distance is short, the flight does not take time to climb to a high altitude.

(6) **Need for surprise.** Surprise is more likely when using low altitude.

(7) **Pilot fatigue.** Terrain flying is more fatiguing for aircrews.

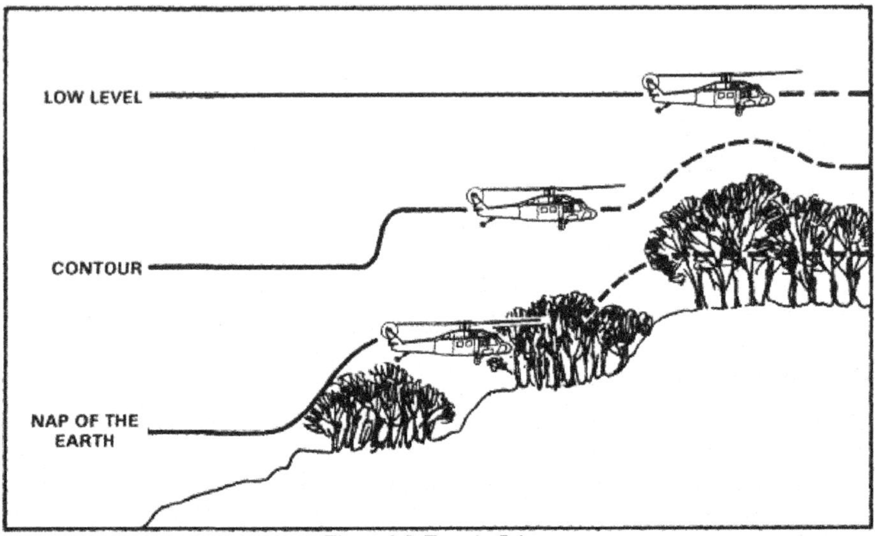

Figure 3-8. Terrain flying.

3-29. Supporting fires along the flight route.

a. Fires along the flight route are planned to suppress known or suspected enemy positions. These fires should be intense and of short duration. Multiple target engagement techniques should be utilized (groups, series).

b. Fire plans cover the PZs, flight routes, and LZs. Fire support plans include suppression of enemy air defense systems and smoke to protect formations from enemy detection. This requires aggressive fire planning by the fire support officer and direct coordination with FA and mortar fire direction centers and other fire support elements.

c. All available fire support is used to suppress and/or destroy enemy weapons including TACAIR, artillery, and attack helicopters.

d. Support may consist of smoke, chaff (air-dropped, shredded aluminum foil to foul radar), or other countermeasures for suppressing or confusing enemy air defense systems.

e. On-call fires are planned along the flight route to ensure rapid adjustment on targets of opportunity.

f. During night operations, the use of illumination fire requires detailed planning. Illumination can interfere with night vision goggles and cause unsafe conditions.

3-30. Air movement timing.

a. A successful air assault operation is a sequence of actions carefully planned and precisely executed.

b. The basis for timing is the time when the first aircraft in the first lift of the operation is to touch down on the LZ. It is referred to as H-hour. All times in air assault operations are referenced from H-hour (landing time column, air movement table). The H-hour in air assault operations is equivalent to the attack time in a mission order. If delays are encountered due to weather or aircraft delays, the commander announces a new H-hour (Figure 3-9).

ANNEX B (AIR MOVEMENT TABLE) to OPORD 6, TF 1-67
Time Zone Used Throughout the Order: ALPHA

(Classification)

Copy No_____ of _____ Copies
Hqs, 1 Bn, 67th Inf
101600 Jan 8__

P - Primary
A - Alternate

AVN UNIT	LIFTED UNIT	LIFT	SERIAL	LOADS	PICKUP ZONE	LOADING TIME	LIFT-OFF TIME	SP TIME	RP TIME	LANDING ZONE	LANDING TIME	REMARKS
	A	1	1	1-4	ALPHA S/L ALTN BRAVO					GREEN S/L ALTN YELLOW	H-HOUR	BUMP 4, 7, 10, 15 FLT RT P-EAGLE A-HAWK
			2	5-8								
			3	9-12	BRAVO S/L					GREEN S/L		
			4	13-16								
	B	2	1	1-4	CHARLIE S/R					GREEN S/R		BUMP 4, 8, 12, 16 FLT RT P-HAWK A-BAT
			2	5-8								
			3	9-12								
			4	13-16								

2 NOW REFER TO CODE-NUMBERED AIR MOVEMENT TABLE (PAGE 3-37) TO SEE HOW THIS INFORMATION APPEARS IN ITS ENTIRETY

Figure 3-9. Landing time.

c. Normal distance from RP to LZ is three to five kilometers. Planning time for navigating this distance is approximately two minutes, depending upon air speed (Figure 3-10). In order for the first aircraft to land in the LZ at H-hour, it must reach the RP at H-2 minutes (RP time column, air movement table).

ANNEX B (AIR MOVEMENT TABLE) to OPORD 6, TF 1-67 (Classification) Copy No ______ of ______ Copies
Time Zone Used Throughout the Order: ALPHA Hqs, 1 Bn, 67th Inf
101500 Jan 8 __

P - Primary
A - Alternate

AVN UNIT	LIFTED UNIT	LIFT	SERIAL	LOADS	PICKUP ZONE	LOADING TIME	LIFT-OFF TIME	SP TIME	RP TIME	LANDING ZONE	LANDING TIME	REMARKS
	A	1	1	1-4	ALPHA S/L ALTN BRAVO				H-2	GREEN S/L ALTN YELLOW		BUMP 4, 7, 10, 15 FLT RT P-EAGLE A-HAWK
			2	5-8								
			3	9-12	BRAVO S/L					GREEN S/L		
			4	13-16								
	B	2	1	1-4	CHARLIE S/R					GREEN S/R		BUMP 4, 8, 12, 16 FLT RT P-HAWK A-BAT
			2	5-8								
			3	9-12								
			4	13-16								

3 NOW REFER TO CODE-NUMBERED AIR MOVEMENT TABLE (PAGE 3-37) TO SEE HOW THIS INFORMATION APPEARS IN ITS ENTIRETY

Figure 3-10. Release point time.

d. The air movement table requires time for detailed planning. For large operations, an LO from the supporting aviation element is required.

e. The importance of an air movement table should not be underemphasized. As stated previously, C2 procedures should be planned to allow continued execution despite loss of radio communications. If the AMC and lift flight leaders have air movement tables in their possession, they can continue the mission without radio communications.

(1) **Flight time.** The following example explains how to compute the time required to cover the distance from the SP to the RP. These times are computed for the entire length of the flight route from the SP to the RP. The length of each of the flight routes is measured so that en route times can be computed.

(2) **Round up** to **next higher whole number.** Figure 3-11 is an example of how the formula is to be used:

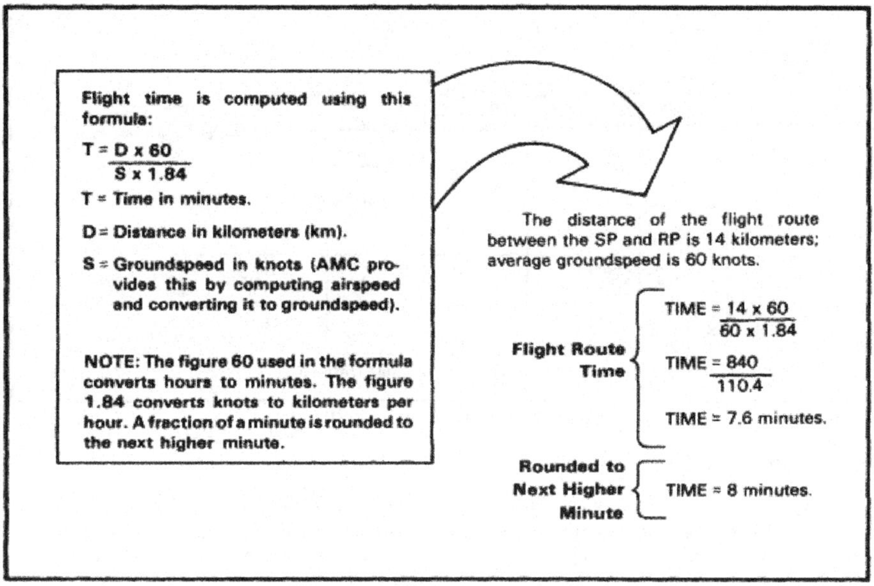

Figure 3-11. Flight time computed.

3-31. Lift-off time, start point time.

a. Lift-off time must be determined first in order to meet the LZ time. The total flight route time is determined by adding the time to fly from the SP to the RP to the LZ. In the preceding example, flight route time was determined to be eight minutes. The flight time from the RP to the LZ is two minutes. In this case, SP time would be H-10 arrived at as follows:

RP time = H-2 (two minutes for RP to LZ).

SP time = H-10 (eight minutes flight route time plus two minutes from RP to LZ).

b. To determine the lift-off time, add the time between PZ and SP. Continuing the example, if time between PZ and SP is two minutes, the -

RP time = H2.

SP time = H-10.

Lift-off time = H-12 (this includes the two minutes from PZ to SP).

c. All times (lift-off, SP, and landing) are recorded in the proper columns of theair movement table (Figure 3-12).

ANNEX B (AIR MOVEMENT TABLE) to OPORD 8, TF 1-67 (Classification) Copy No_____ of _____ Copies
Time Zone Used Throughout the Order: ALPHA Hqs. 1 Bn, 67th Inf
P - Primary 101500 Jan 8__
A - Alternate

AVN UNIT	LIFTED UNIT	LIFT	SERIAL	LOADS	PICKUP ZONE	LOADING TIME	LIFT-OFF TIME	SP TIME	RP TIME	LANDING ZONE	LANDING TIME	REMARKS
	A	1	1	1-4	ALPHA S/L ALTN BRAVO		H-12	H-10	H-2	GREEN S/L ALTN YELLOW	H-HOUR	BUMP 4,7,10,15 FLT RT P-EAGLE A-HAWK
			2	5-8								
			3	9-12	BRAVO S/L					GREEN S/L		
			4	13-16								
	B	2	1	1-4	CHARLIE S/R					GREEN S/R		BUMP 4,8,12,16 FLT RT P-HAWK A-BAT
			2	5-8								
			3	9-12								
			4	15-16								

4 NOW REFER TO CODE-NUMBERED AIR MOVEMENT TABLE (PAGE 3-37) TO SEE HOW THIS INFORMATION APPEARS IN ITS ENTIRETY

Figure 3-12. Lift-off time and SP time.

3-32. Loading time.

a. Loading time is the time required, before lift-off, to load the aircraft. Time to load is normally dependent on prior training, equipment to be carried, and light conditions. Night operations require more loading time. Once loading time is determined, it is added to the previously computed times (Figure 3-13).

ANNEX B (AIR MOVEMENT TABLE) to OPORD 8, TF 1-67 (Classification) Copy No_____ of _____ Copies
Time Zone Used Throughout the Order: ALPHA Hqs. 1 Bn, 67th Inf
P - Primary 101500 Jan 8__
A - Alternate

AVN UNIT	LIFTED UNIT	LIFT	SERIAL	LOADS	PICKUP ZONE	LOADING TIME	LIFT-OFF TIME	SP TIME	RP TIME	LANDING ZONE	LANDING TIME	REMARKS
	A	1	1	1-4	ALPHA S/L ALTN BRAVO	H-16	H-12	H-10	H-2	GREEN S/L ALTN YELLOW	H-HOUR	BUMP 4,7,10,15 FLT RT P-EAGLE A-HAWK
			2	5-8		H-15	H-11	H-9	H-1		H+1	
			3	9-12	BRAVO S/L	H-14	H-10	H-8	H	GREEN S/L	H+2	
			4	13-16		H-13	H-9	H-7	H+1		H+3	
	B	2	1	1-4	CHARLIE S/R					GREEN S/R		BUMP 4,8,12,16 FLT RT P-HAWK A-BAT
			2	5-8								
			3	9-12								
			4	15-16								

5 NOW REFER TO CODE-NUMBERED AIR MOVEMENT TABLE (PAGE 3-37) TO SEE HOW THIS INFORMATION APPEARS IN ITS ENTIRETY.

Figure 3-13. Loading time.

b. If a unit requires four minutes to load, in order to meet the lift-off time of H-12, it begins loading no later than H-16 minutes (loading time and lift-off time columns, air movement table).

c. Slingloading also requires additional time that must be calculated into this plan. The use of slingloads may also reduce the en route airspeed.

d. The aviation and ground elements each begin movement to the PZ to start loading at the prescribed time.

e. With the air movement time schedule completed for the initial elements, the air movement table is completed by building on the initial times.

3-33. Loading time, second lift.

The first lift is planned working back from H-hour. Planning for the second lift is forward from H-hour since the en route times are established (Figure 3-14). However, before annotating the air movement time for the second element, the time required for the aircraft to return from the LZ to the PZ (along the return route) is computed. This is done employing the same method used in computing initial flight time. For example, if it is two minutes flight time from the PZ to the SP, two minutes from the RP to the LZ, and eight minutes en route, then once the aircraft have unloaded, it takes 12 minutes to return to the PZ (lift, serial, loads, and loading time columns, air movement table).

ANNEX B (AIR MOVEMENT TABLE) to OPORD 6, TF 1-67
Time Zone Used Throughout the Order: ALPHA
P - Primary
A - Alternate

(Classification)

Copy No____ of ____ Copies
Hqs. 1 Bn. 67th Inf
101500 Jan 8__

AVN UNIT	LIFTED UNIT	LIFT	SERIAL	LOADS	PICKUP ZONE	LOADING TIME	LIFT-OFF TIME	SP TIME	RP TIME	LANDING ZONE	LANDING TIME	REMARKS
	A	1	1	1-4	ALPHA S/L ALTN BRAVO	H-16	H-12	H-10	H-2	GREEN S/L ALTN YELLOW	H-HOUR	BUMP 4, 7, 10, 15 FLT RT P-EAGLE A-HAWK
			2	5-8		H-15	H-11	H-9	H-1		H+1	
			3	9-12	BRAVO S/L	H-14	H-10	H-8	H	GREEN S/L	H+2	
			4	13-16		H-13	H-9	H-7	H+1		H+3	
	B	2	1 2	1-4 5-8	CHARLIE S/R	H+12						

6 NOW REFER TO CODE-NUMBERED AIR MOVEMENT TABLE (PAGE 3-37) TO SEE HOW THIS INFORMATION APPEARS IN ITS ENTIRETY.

Figure 3-14. Load time, second lift.

3-34. Second lift.

If it takes four minutes to load the second lift, then the lift-off time is recorded as H+16 minutes. Subsequently, if it takes two minutes to get from PZ to SP, SP time is H+18 minutes. If the flight requires 10 minutes to get from SP to RP, RP time is H+28 minutes. Once again, allow two minutes for transition from RP to the LZ; the landing time for the second element on its LZ is H+30 minutes. Subsequent times are all computed in the same manner (landing time column, air movement table) (Figure 3-15).

ANNEX B (AIR MOVEMENT TABLE) to OPORD 6, TF 1-67
Time Zone Used Throughout the Order: ALPHA
P - Primary
A - Alternate

(Classification)

Copy No ______ of ______ Copies
Hqs. 1 Bn, 67th Inf
101500 Jan 8 __

AVN UNIT	LIFTED UNIT	LIFT	SERIAL	LOADS	PICKUP ZONE	LOADING TIME	LIFT-OFF TIME	SP TIME	RP TIME	LANDING ZONE	LANDING TIME	REMARKS
	A	1	1	1-4	ALPHA S/L ALTN BRAVO	H-16	H-12	H-10	H-2	GREEN S/L ALTN YELLOW	H-HOUR	BUMP 4, 7, 10, 15 FLT RT P-EAGLE A-HAWK
			2	5-8		H-15	H-11	H-9	H-1		H+1	
			3	9-12	BRAVO S/L	H-14	H-10	H-8	H	GREEN S/L	H+2	
			4	13-16		H-13	H-9	H-7	H+1		H+3	
	B	2	1	1-4	CHARLIE S/R	H+12	H+16	H+18	H+28	GREEN S/R	H+30	BUMP 4, 8, 12, 16 FLT RT P-HAWK A-BAT
			2	5-8		H+13	H+17	H+19	H+29		H+31	
			3	9-12		H+14	H+18	H+20	H+30		H+32	
			4	13-16		H+15	H+19	H+21	H+31		H+33	

7 NOW REFER TO CODE-NUMBERED AIR MOVEMENT TABLE (PAGE 3-37) TO SEE HOW THIS INFORMATION APPEARS IN ITS ENTIRETY

Figure 3-15. Second lift.

3-35. Planning for refueling.

a. An accurate table must also contain the times involved in aircraft refueling.

b. Refueling is planned so that a flight completes refueling before the serial that is scheduled to be refueled last gets critically low on fuel. If only a portion of the flight can be refueled at the FARP, the first serial might have to refuel as much as an hour before it needs refueling. The other serials continue the lift operation until it is their turn to refuel. The plan should allow a smooth, continuous rotation of aircraft into and out of the FARP (Figure 3-16).

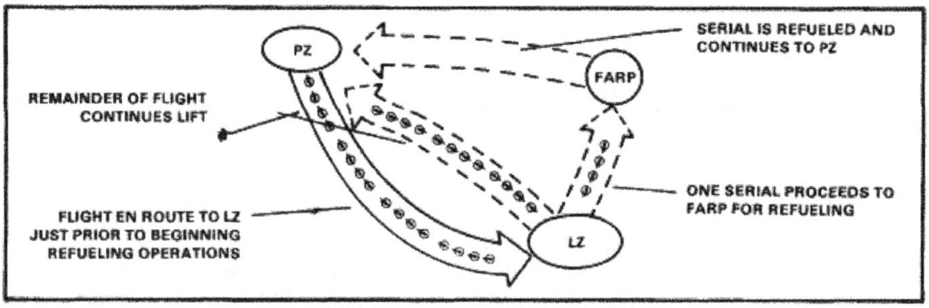

Figure 3-16. Sequencing serials into the FARP.

c. The number of aircraft that can refuel at one time is divided into the number of aircraft in the lift. The result is the number of separate trips to the FARP to refuel the entire lift one time.

d. The result is multiplied by the time required for the aircraft to refuel. The answer gives the total time required to refuel one time. For example, if four serials require 15 minutes each to refuel, it takes one hour to refuel the entire lift. (This time includes time for repositioning to the PZ.)

e. This figure is subtracted from the available flying time to determine at what point in the mission the refueling process must begin.

f. The physical planning for refueling begins when the aircraft arrive in the PZ for loading. Once the plans are developed, refueling becomes a part of the table (Figure 3-17).

Section VI

LOADING PLAN

3-36. General.

The loading plan is based on the air movement plan. It ensures that troops, equipment, and supplies are loaded on the correct aircraft. Unit integrity is maintained when aircraft loads are planned. However, assault forces and equipment may be cross-loaded so that command and control assets, all types of combat power, and a mix of weapons arrive at the LZ ready to fight. Aircraft loads are also placed in priority to establish a bump plan. A bump plan ensures that essential troops and equipment are loaded ahead of less critical loads in case of aircraft breakdown or other problems. Below brigade level, loading plans are established by SOPS. In any case, planning must cover the organization and operation of the PZ including load positions, day and night markings, and communications. 'he loading plan is most important when mixing internal and external loads and/or when mixing aircraft types (lift and medium helicopters).

3-37. Coordination with air mission control.

a. Loading plans are carefully coordinated with the AMC or the aviation LO. Copies of the loading plan should be distributed to the aviation LO, command and control elements, AMC, and the PZ control officer.

b. For battalion or larger air assault operations, a written plan may be required to:

(1) Control movement of troops, supplies, and equipment to and about the PZ.

(2) Designate unit loading sites.

(3) Control timing for arrival, loading, and departing of aircraft.

c. The requirement for detailed, written plans can be reduced by having adequate unit SOPs covering PZ operations and loading plans. Regardless of SOP adequacy, the loading phase should receive command attention to ensure that it goes smoothly. A well-planned and properly executed loading operation is imperative to mission success.

ANNEX B (AIR MOVEMENT TABLE) to OPORD 6, TF 1-67

Time Zone Used Throughout the Order: ALPHA

(Classification)

Copy No ______ of ______ Copies
Hqs, 1 Bn, 67th Inf
101500 Jan 8 ___

P - Primary
A - Alternate

AVN UNIT	LIFTED UNIT	LIFT	SERIAL	LOADS	PICKUP ZONE	LOADING TIME	LIFT-OFF TIME	SP TIME	RP TIME	LANDING ZONE	LANDING TIME	REMARKS
	A	1	1	1-4	ALPHA S/L ALTN BRAVO	H-16	H-12	H-10	H-2	GREEN S/L ALTN YELLOW	H-HOUR	BUMP 4, 7, 10, 15 FLT RT P-EAGLE A-HAWK
			2	5-8		H-15	H-11	H-9	H-1		H+1	
			3	9-12	BRAVO S/L	H-14	H-10	H-8	H	GREEN S/L	H+2	
			4	13-16		H-13	H-9	H-7	H+1		H+3	
	B	2	1	1-4	CHARLIE S/K	H+12	H+16	H+18	H+28	GREEN S/K	H+30	BUMP 4, 8, 12, 16 FLT RT P-HAWK A-BAT
			2	5-8		H+13	H+17	H+19	H+29		H+31	
			3	9-12		H+14	H+18	H+20	H+30		H+32	
			4	13-16		H+15	H+19	H+21	H+31		H+33	
		2	1	1-4	FARP#1	H+40	REFUEL					FARP LOCATION-AS 382 641
	DSC	3	1	5-8	DELTA	H+46	H+49	H+52	H+60	PURPLE	H+62	BUMP 5, 8, 10 FLT RT P-OWL A-FALCON
	CSC		2	9-12		H+47	H+50	H+53	H+61	ALTN WHITE	H+63	
	B		3	13-16		H+48	H+51	H+54	H+62		H+64	
	B		4	1-4		H+55	H+58	H+61	H+69		H+71	
		3	1	5-8	FARP#1	H+77	REFUEL					
	C	4	1	9-12	CHARLIE S/L	H+83	H+86	H+88	H+98	WHITE S/L ALTN ORANGE	H+100	RELEASE TWO AIRCRAFT FLT RT P-HAWK A-CROW BUMP 4, 8
			2	13-16		H+84	H+87	H+89	H+99		H+101	
			3	1-4		H+91	H+94	H+96	H+106		H+108	
			4	5-8		H+92	H+95	H+97	H+107		H+109	
		4	1	9-12	FARP#1	H+115	REFUEL					
		5	1	15-16	CHARLIE S/L	H+117	H+120	H+122	H+132		H+134	BUMP 15
	A		2	1-4	FOXTROT S/K	H+118	H+121	H+123	H+133	AMBER S/K	H+135	BUMP 6, 9 FLT RT P-PHOENIX A-ROBIN
	C		3	5-8		H+119	H+122	H+124	H+134		H+136	
	B		4	9-10		H+130	H+133	H+135	H+145		H+147	BUMP 10
		5	1	13-16	FARP#1	H+149	REFUEL					

(Classification)

8 NOW REFER TO CODE-NUMBERED AIR MOVEMENT TABLE (PAGE 3-37) TO SEE HOW THIS INFORMATION APPEARS IN ITS ENTIRETY

Figure 3-17. Refueling.

3-38. Pickup zone selection.

a. Pickup zone identification is the first step in the loading plan development.

NOTES: 1. Alternate PZs are identified at the same time. The goal of PZ identification is to locate suitable areas to accommodate the lift aircraft.

2. The specifications (such as degree of slope, wind speeds, and distance between aircraft) used in this chapter to identify and select PZs, LZs, and flight routes, are planning guides for the ground unit. They may be adjusted by the AMC based on his evaluation of his unit's level of training.

b. Extraction under pressure is also a function of successful PZ selection. Selected PZs must facilitate delivery of suppressive fires and continuous security of elements to be extracted and the helicopters.

c. Figure 3-18 shows the technical factors that must be considered in selecting Pzs.

SIZE

PZ and LZ size requirements depend on type and number of aircraft and are based on minimum acceptable distances between aircraft. Each aircraft should be provided a circular landing point separated from other aircraft and free of obstacles. Minimum recommended landing point sizes (diameter of circle in meters) are:

(1) Observation helicopters--25 meters.

(2) UH-1, AH-1--35 meters.

(3) UH-60, AH-64--50 meters.

(4) Cargo helicopters--80 meters.

SURFACE CONDITIONS

Surface conditions in the PZ and LZ should not conceal the touchdown point or create hazards to landing (e.g., sand, blowing dust, snow). The surface of the zone should be free of obstacles that could damage landing aircraft (no tree stumps, large rocks). It must be firm enough to support the traffic. Drainage should be adequate for rainfall runoff. If the surface is contaminated (chemical or radiological) to an unacceptable degree, it may preclude use of the area. If part of an area is unsatisfactory for any reason, that part is not used.

GROUND SLOPE — LANDING

As a guide: if the ground slope is 0 to 6 percent, land upslope; if the slope is 7 to 15 percent, land sideslope; over 15 percent, no touchdown (aircraft may hover to drop off or pick up personnel and/or equipment).

IF SLOPE IS 0% - 6%, LAND UPSLOPE.

IF SLOPE IS 7% - 15%, LAND SIDESLOPE.

NOTE: EXIT ON DOWNHILL SIDE TO AVOID ROTOR BLADES.

Figure 3-18. Technical factors in PZ and LZ selections.

OBSTACLES

For planning purposes, an obstacle clearance ratio of 10 to 1 is used on the approach and departure ends of the PZ and LZ. That is, a landing point requires 100 feet of horizontal clearance if a helicopter must approach or depart directly over a 10-foot tall tree.

A lesser ratio may be used if the helicopter executes a steep approach or departure in emergency situations or with light loads.

All obstacles within the PZ and LZ are marked with red lights at night (turned on only when PZ or LZ is in use), or red panels during the day. The markings are not used if they cause the position to be seen by the enemy.

APPROACH/DEPARTURE

The terrain surrounding a possible PZ or LZ is analyzed for air traffic patterns. In a tactical situation, constantly approaching the PZ or LZ over the same ground track should be avoided. Still, there are only so many ways to get into an area. Approaches should be free of obstacles, and landings should be made into the wind. Ideally, approach and departure are made along the long axis of the LZ over the lowest obstacle, and into the wind.

LOADS

When a helicopter is loaded to near maximum lift capacity, it requires longer distances to lift-off and land (it cannot ascend or descend vertically). The greater the load (near or at maximum), the larger the PZ and LZ must be to accommodate a flight.

Figure 3-18 Continued.

d. Once available PZs are identified, the AATFC and his S3 select and assign PZs to be used by each unit. Pickup zone criteria include:

(1) **Number.** Multiple PZs avoid concentrating forces in one area.

(2) **Size.** Each PZ should accommodate all supporting aircraft at once, if possible.

(3) **Proximity** to **troops.** Where possible, the selected PZs should not require extensive ground movement (to the PZ) by troops.

(4) **Accessibility.** Pickup zones should be accessible to vehicles to move support assets and infantry.

(5) **Vulnerability to attack.** Selected PZs should be masked by terrain from enemy observation.

(6) **Preparation.** It is better to select PZs that are usable as they are, or require only limited clearing (Figure 3-19).

Each unit is assigned a primary and alternate PZ. The alternate PZ is used if (for any reason) the primary cannot be used. The unit is also assigned an LZ and an alternate LZ. This information is annotated on the air movement table and PZ/LZ overlay and given to subordinate units.

ANNEX B (AIR MOVEMENT TABLE) to OPORD 6, TF 1-67
Time Zone Used Throughout the Order: ALPHA

(Classification)

Copy No______ of ______ Copies
Hqs, 1 Bn, 67th Inf
101600 Jan 8 __

P - Primary
A - Alternate

AVN UNIT	LIFTED UNIT	LIFT	SERIAL	LOADS	PICKUP ZONE	LOADING TIME	LIFT-OFF TIME	SP TIME	RP TIME	LANDING ZONE	LANDING TIME	REMARKS
					ALPHA					GREEN		
					ALTN BRAVO					ALTN YELLOW		

[9] NOW REFER TO CODE-NUMBERED AIR MOVEMENT TABLE (PAGE 3-37) TO SEE HOW THIS INFORMATION APPEARS IN ITS ENTIRETY.

Figure 3-19. Pick-up and landing zones.

3-39. Pickup zone control officer.

a. The pickup zone control officer organizes, controls, and coordinates operationsin PZs selected by the AATFC. (The S4 selects and controls logistical PZs.)

b. The PZCO accomplishes the following:

(1) **Forms control group.** To manage operations, the PZCO forms a control group to assist him. It may include air traffic control, subordinate units, and support personnel (manpower to clear the PZ; security). The PZCO selects a central location to position the group. The PZCO is designated by the AATFC, usually the S3 Air. For battalion air assault operations, each company commander appoints a PZCO who operates a company PZ for the battalion.

(2) **Establishes communications.** The PZCO should communicate on two primary radio frequencies: one to control movement and loading of units, and one to control aviation elements (combat aviation net). Alternate frequencies are provided as necessary.

(3) **Plans and initiates fire support.** He plans fires near PZs to provide all-round protection (from available support) without endangering arrival and departure of troops or aircraft.

(4) **Plans and initiates security.** The PZCO ensures that adequate security is provided. Security protects the main body as it assembles, moves to the PZ, and is lifted out. Security elements should be provided by other forces if the PZ is within a friendly area. Security elements should be provided by other forces if the PZ is within a friendly area. Security comes from AATF resources if it is to be extracted from the objective area.

(5) **Clears PZ of obstacles.**

(6) **Marks the PZ.**

3-40. Pickup zone marking.

a. The PZCO directs the marking of PZs. An effective method is to name the PZ by color and mark it accordingly (by that color) to direct where aircraft will land.

b. Red is never used to mark an aircraft landing position. It is used to mark landing obstacles such as trees or stumps in the landing area.

c. Regardless of the type of markers, the PZ is marked to indicate where aircraft are to land and coincides with the selected PZ aircraft formation.

d. An effective method is to have several individuals in each unit paint (and carry) an extra camouflage cover or a modified (cut to size) VS-17 panel. The colored covers, when displayed, indicate where the lead aircraft lands.

3-41. Movement to pickup zone.

a. Ground and aviation unit movement to the PZ is scheduled so that only the troops to load, and the helicopter to be loaded, arrive at the PZ at the same time. This prevents congestion, preserves security, and reduces vulnerability to enemy actions on the PZ (Figure 3-20).

The PZCO, working with the AMC, selects and coordinates aircraft landing formations and approach headings. This information is entered on the air movement table. Flight formations in the PZs are used to facilitate loading. They may be dictated by the number and type aircraft and the configuration and size of the PZs. (Aircraft formations are standard, but may be modified to meet deployment requirements.) If possible, PZ formations should be the same as the formation that will be used in the LZ. The air movement table below reflects a staggered trail left (S/L) formation for both the PZ and LZ.

ANNEX B (AIR MOVEMENT TABLE) to OPORD 6, TF 1-67
Time Zone Used Throughout the Order: ALPHA

(Classification)

Copy No_____ of _____ Copies
Hqs, 1 Bn, 67th Inf
101600 Jan 8 __

P - Primary
A - Alternate

AVN UNIT	LIFTED UNIT	LIFT	SERIAL	LOADS	PICKUP ZONE	LOADING TIME	LIFT-OFF TIME	SP TIME	RP TIME	LANDING ZONE	LANDING TIME	REMARKS
					ALPHA S/L ALTN BRAVO					GREEN S/L ALTN YELLOW		

10 NOW REFER TO CODE NUMBERED AIR MOVEMENT TABLE (PAGE 3-37) TO SEE HOW THIS INFORMATION APPEARS IN ITS ENTIRETY

Figure 3-20. Pickup zone and landing sone formations.

b. To coordinate the movement of units to the PZ, the PZCO:

(1) Selects troop assembly areas, holding areas, and routes of movement. A holding area is located close to the PZ. It is used only when the assembly area is some distance away and does not allow timely movement to the PZ.

(2) Determines movement time of ground units to PZ.

(3) Specifies arrival time(s) and sees that movement of units remains on schedule.

3-42. Airloading table.

a. At company and lower levels, the airloading table assigns each man and major items of equipment or supplies to a specific aircraft (chalk).

b. The airloading table is an accountability tool, a loading manifest for each aircraft.

c. When time is limited, the table can be put on a sheet of paper from a squad leader's notebook. It should list soldiers (by name) and equipment to be loaded on each chalk. These lists are left with a specified representative in the area for consolidation or exchange between aircraft troop commanders (senior person in each aircraft load). This procedure ensures that if an aircraft is lost, a list of personnel and equipment on board is available.

3-43. Load planning.

a. During preparation of the loading tables, unit leaders at all levels attempt to maintain the following:

(1) **Tactical integrity of units.** When planning loads for air assault operations, fire teams and squads are loaded intact on the same aircraft, and platoons in the same serial. This ensures integrity as a fighting unit upon landing. The commander's goal is to load his unit so that unit integrity is maintained at every level.

(2) **Self-sufficiency of loads.** Each unit load should be functional by itself (whenever possible).

(a) Every towed item is accompanied by its prime mover.

(b) Crews are loaded with their vehicle or weapon.

(c) Component parts accompany the major items of equipment.

(d) Ammunition is carried with the weapon.

(e) Sufficient personnel are on board to unload cargo carried.

(f) Communication between chalks, if possible, without using the aircraft radios.

(3) **Tactical cross-loading.** Loads should be planned so that all leaders, or all crew-served weapons, are not loaded on the same aircraft. Thus, if an aircraft is lost, the mission is not seriously hampered. For example, loading the platoon leader, platoon sergeant, and all the squad leaders on the same helicopter, or loading more than one machine gun team on the same aircraft, are violations of cross-loading principles.

b. Another consideration is to determine whether internal or external loading is the best delivery method for equipment and supplies. Helicopters loaded internally can fly faster and are more maneuverable. Externally (sling) loaded helicopters fly slower and are less maneuverable; however, they can be loaded and unloaded more rapidly than internally loaded helicopters. The method used depends largely on availability of sling and rigging equipment.

NOTE: Supplies loaded externally (although loaded rapidly) can present problems if the supplies are destined for more

than one location or unit.

3-44. Aircraft bump plan.

a. Each aircraft load has a bump sequence designated on its airloading table. Bump priority ensures that the most essential personnel and equipment arrive at the objective area first. It specifies personnel and equipment that may be bumped and delivered later.

b. If all personnel within the load cannot be lifted, individuals must know who is to offload and in what sequence. This ensures that key personnel are not bumped arbitrarily. Also, bump sequence is designated for aircraft within each serial or flight. This sequence is listed on the air movement table.

c. This also ensures that key aircraft loads are not left in the PZ. When an aircraft within a serial or flight cannot lift off, and key personnel are on board, they offload and reboard another aircraft that has priority.

3-45. Aircraft bump-and-straggler control.

A PZ bump-and-straggler collection point is specified by company or larger units. Personnel not moved as planned report to this location, are accounted for, regrouped, and rescheduled by the PZCO for later delivery to appropriate LZs.

3-46. Lifts, serials, and loads.

To maximize operational control, aviation assets are designed into lifts, serials, and loads (Figure 3-21).

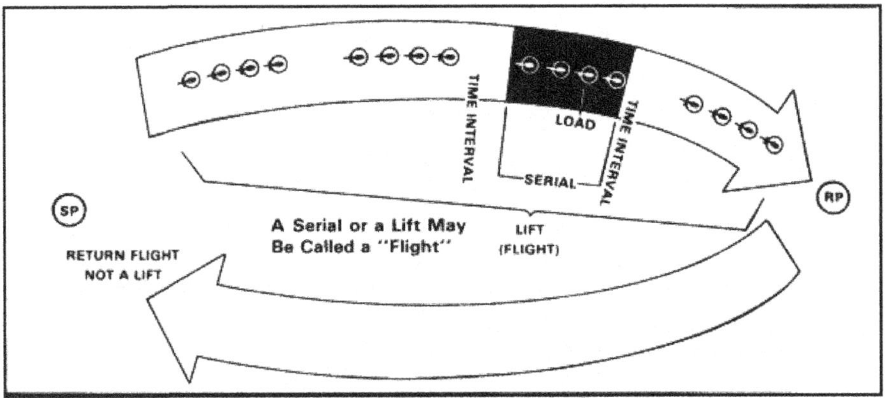

Figure 3-21. Lifts, serials, and loads.

a. **Lifts.** A lift is one sortie of all utility and cargo aircraft assigned to a mission. That is, each time all assigned aircraft pick up troops and/or equipment and set them down on the LZ, one lift is completed. The second lift is completed when all lift aircraft place their second load on the LZ.

b. **Serials.**

(1) There may be times when a lift is too large to fly in one formation. In such cases, the lift is organized into a number of serials. A serial is a tactical grouping of two or more aircraft under the control of a serial commander (aviator), and separated from other tactical groupings within the lift by time or space. The use of serials may be necessary to maintain effective control of aviation assets. For example, if NOE flight is used, it would be difficult to control 16 aircraft as a single increment; however, a 16-aircraft lift with four serials of four aircraft each could be more easily controlled.

(2) Serials may also be required when the capacity of available PZs or LZs is limited. If there is a lift of 16 aircraft and available PZs and/or LZs will accommodate only four aircraft, it is best to organize into four serials of four aircraft each.

(3) Serials are employed to take advantage of available flight routes. If there are several acceptable flight routes, the AATFC may choose to employ serials to avoid concentrating his force along one flight route. If the commander wants all his forces to land simultaneously in a single LZ, he does so by having the serials converge at a common RP before landing. With a lift of 16 aircraft and four available flight routes, the AATFC could use four serials of four aircraft each. Each serial would use a different flight route. Each time there is a new lift, a new serial begins. For example, in lift one, there are serials one through four; in lift two, serials again start with one.

c. **Loads.**

(1) Within each lift, there is also a specific number of loads. A load is personnel and/or equipment designated to be moved by a specific aircraft. When planning the air movement, each aircraft within the lift is termed a load. For example, within a lift of 10, there are aircraft loads one through 10. For each lift thereafter, there will also be loads one through 10. Each aircraft is accounted for within each lift.

(2) An aircraft load may also be referred to as a chalk load," "chalk number," or a "chalk." Loads also must be designated within serials just as they are within lifts. Counting within the serials is continuous up to the total number of aircraft in the lift. For example, in a lift of 16 aircraft in lift one, serial one, there may be loads one through four. In lift one, serial two, there may be loads five through eight. In lift one, serial three, there may be loads nine through 12. Finally, in lift one, serial four, there may be loads 13 through 16 (Figure 3-22).

These three elements on the air movement table (lift, serial, load) collectively designate the specific aircraft that are used to move a ground unit. A term that is inclusive of lift and serial is a flight. A flight is two or more aircraft directed by a single aviator or a flight leader. Therefore, a lift may be referred to as a flight, as may a serial. "Flight" is not used in an air movement document; it is used in referring to aviation resources (a flight of four).

ANNEX B (AIR MOVEMENT TABLE) to OPORD 6, TF 1-67 (Classification) Copy No_____ of _____ Copies

Time Zone Used Throughout the Order: ALPHA

Hqs, 1 Bn, 67th Inf

101500 Jan 8 __

P - Primary

A - Alternate

AVN UNIT	LIFTED UNIT	LIFT	SERIAL	LOADS	PICKUP ZONE	LOADING TIME	LIFT-OFF TIME	SP TIME	RP TIME	LANDING ZONE	LANDING TIME	REMARKS
		1	1	1-4	ALPHA S/L ALTN BRAVO					GREEN S/L ALTN YELLOW		BUMP 4, 7, 10, 15
			2	5-8								
			3	9-12	BRAVO S/L					GREEN S/L		
			4	13-16								
		2	1	1-4	CHARLIE S/R					GREEN S/R		BUMP 4, 8, 12, 16
			2	5-8								
			3	9-12								
			4	13-16								

[11] NOW REFER TO CODE-NUMBERED AIR MOVEMENT TABLE (PAGE 3-37) TO SEE HOW THIS INFORMATION APPEARS IN ITS ENTIRETY

*NOTE: Aircraft bump sequence is shown under the remarks section of the air movement table. The aircraft to be bumped first is listed first. Other aircraft to be bumped, if necessary, are listed in the sequence they are to be bumped. For example: the remarks section reflects BUMP 4, 7, 10, and 15. If one aircraft load has to be bumped, it will be load 4. If four loads have to be bumped, loads 4, 7, 10, and 15 (in that order) will be bumped.

Figure 3-22. Aircraft load information.

3-47. Sequence of departure.

The sequence of departure from PZs is based on the mission to be accomplished by each subordinate unit upon landing. Unit priorities are based on the sequence of arrival at their LZs. Units are scheduled to depart (in order) based on en route time to the LZ. For example, if Company A is to land first (at H-hour), and Company B second (at H+5), and Company B is 15 minutes farther (in flight time) from the LZ, it may depart the PZ before Company A.

Section VII

STAGING PLAN

3-48. General.

The staging plan is based on the loading plan and prescribes the arrival time of ground units (troops, equipment, and supplies) at the PZ in the proper order for movement.

3-49. Procedures.

Loads must be ready before aircraft arrive at the PZ; usually, ground units are expected to be in PZ posture 15 minutes before aircraft arrive. The staging plan also restates the PZ organization, defines flight routes to the PZ, and provides instructions for linkup of all aviation elements. Air-to-air linkup of aviation units should be avoided, especially at night when night vision goggles are being used.

Section VIII

AIR MISSION BRIEFING

3-50. General.

The air mission briefing (AMB) is the last coordination meeting of key participants in an air assault mission and ensures that key aviation personnel are briefed and that the details of each plan are finalized.

3-51. Content.

The briefing covers the details of all planning and preparation. It covers the operation from beginning to end, which includes the five plans explained previously, and the completed air movement table (Figure 3-23).

3-52. Participants.

The AMB should include, as a minimum, the S2, S3, fire support officer, aviation liaison officer (LO), ADA liaisons, aviation unit operations officer, AMC, battle team captains (BTC) from air reconnaissance and m attack helicopter units, and the ground tactical commander of the unit being inserted. It may also include the S3 Air, communication-electronics officer (CEO), and the S4.

3-53. Location.

The briefing is conducted at a location specified by the AATFC and is dependent upon the amount of time available before execution of the mission. In determining the location, factors to consider are:

- Time required to assemble key personnel.

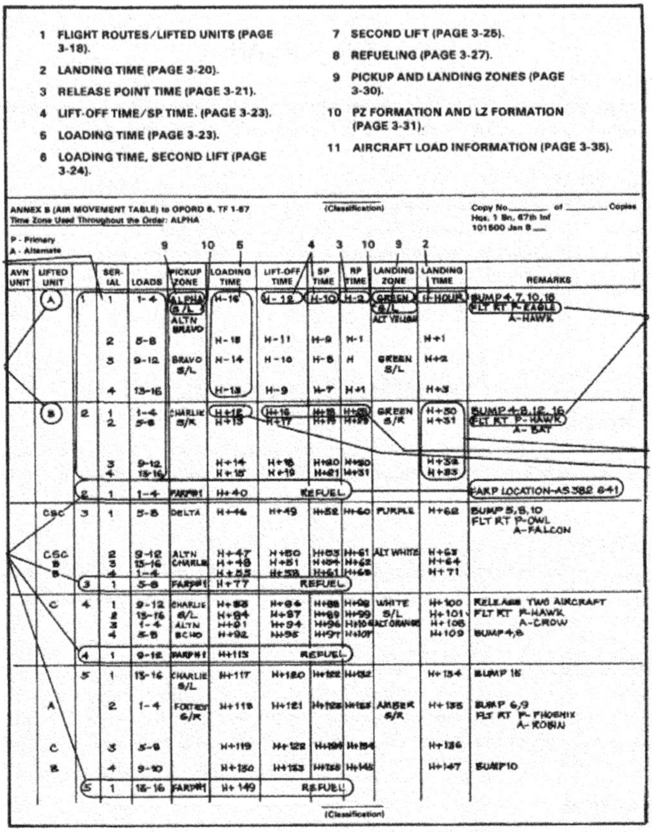

1 FLIGHT ROUTES/LIFTED UNITS (PAGE 3-18).
2 LANDING TIME (PAGE 3-20).
3 RELEASE POINT TIME (PAGE 3-21).
4 LIFT-OFF TIME/SP TIME. (PAGE 3-23).
5 LOADING TIME (PAGE 3-23).
6 LOADING TIME, SECOND LIFT (PAGE 3-24).
7 SECOND LIFT (PAGE 3-25).
8 REFUELING (PAGE 3-27).
9 PICKUP AND LANDING ZONES (PAGE 3-30).
10 PZ FORMATION AND LZ FORMATION (PAGE 3-31).
11 AIRCRAFT LOAD INFORMATION (PAGE 3-35).

ANNEX B (AIR MOVEMENT TABLE) to OPORD 6, TF 1-87
Time Zone Used Throughout the Order: ALPHA

(Classification)

Copy No_____ of _____ Copies
Hqs, 1 Bn, 87th Inf
101500 Jan 8__

P - Primary
A - Alternate

AVN UNIT	LIFTED UNIT		SERIAL	LOADS	PICKUP ZONE	LOADING TIME	LIFT-OFF TIME	SP TIME	RP TIME	LANDING ZONE	LANDING TIME	REMARKS
	A	1	1	1-4	ALPHA S/L ALTN BRAVO	H-16	H-12	H-10	H-2	GREEN S/L ALT YELLOW	H-HOUR	BUMP 4, 7, 10, 16 FLT RT P-EAGLE A-HAWK
			2	5-8		H-15	H-11	H-9	H-1		H+1	
			3	9-12	BRAVO S/L	H-14	H-10	H-8	H	GREEN S/L	H+2	
			4	13-16		H-13	H-9	H-7	H+1		H+3	
	B	2	1 2	1-4 5-8	CHARLIE S/R	H+12 H+13	H+16 H+17	H+18 H+19	H+28 H+29	GREEN S/R	H+30 H+31	BUMP 4, 8, 12, 16 FLT RT P-HAWK A-BAT
			3 4	9-12 13-16		H+14 H+15	H+18 H+19	H+20 H+21	H+30 H+31		H+32 H+33	
		2	1	1-4	FARP#1	H+40	REFUEL					FARP LOCATION-AS 382 641
	CSC	3	1	5-8	DELTA	H+46	H+49	H+52	H+60	PURPLE	H+62	BUMP 5, 8, 10 FLT RT P-OWL A-FALCON
	CSC B B		2 3 4	9-12 13-16 1-4	ALTN CHARLIE	H+47 H+48 H+55	H+50 H+51 H+58	H+53 H+54 H+61	H+61 H+62 H+69	ALT WHITE	H+63 H+64 H+71	
		3	1	5-8	FARP#1	H+77	REFUEL					
	C	4	1 2 3 4	9-12 13-16 1-4 5-8	CHARLIE S/L ALTN ECHO	H+83 H+84 H+91 H+92	H+86 H+87 H+94 H+95	H+88 H+89 H+96 H+97	H+98 H+99 H+106 H+107	WHITE S/L ALT ORANGE	H+100 H+101 H+108 H+109	RELEASE TWO AIRCRAFT FLT RT P-HAWK A-CROW BUMP 4, 8
		4	1	9-12	FARP#1	H+113	REFUEL					
		5	1	13-16	CHARLIE S/L	H+117	H+120	H+122	H+132		H+134	BUMP 16
	A		2	1-4	FOXTROT S/R	H+118	H+121	H+123	H+133	AMBER S/R	H+135	BUMP 6, 9 FLT RT P-PHOENIX A-ROBIN
	C		3	5-8		H+119	H+122	H+124	H+134		H+136	
	B		4	9-10		H+130	H+133	H+135	H+145		H+147	BUMP 10
		5	1	13-16	FARP#1	H+149	REFUEL					

(Classification)

Figure 3-23. Completed air movement table.

- Availability of information (intelligence and status of the aviation assets). Availability of planners.
- Operations security (multiple aircraft in forward battalion locations).
- Vulnerability to enemy observation and fires.

3-54. Time.

The AMB should be held as soon as possible after the reverse planning sequence is completed. Applying the one-third, two-thirds rule should allow enough time for all element leaders to brief their subordinate units. For example, with a 12-hour warning, the briefing should be conducted eight hours before the first aircraft arrives at the PZ. This allows four hours to plan the operation. When air assault forces are required to conduct hasty assaults, there is little time for planning. Establishing habitual relationships and SOPs reduces planning and preparation time. Aircrew endurance must be considered when establishing a time for the AMB. If the operation will be of long duration, an early AMB may not be advisable.

3-55. Air mission briefing format.

The following format is a guide. Its use will help ensure that essential information is included in air assault mission briefings.

(Classification)

TASK FORCE ORGANIZATION

1 . Situation.

a. Enemy forces (especially troop concentrations and locations and types of ADA assets).

b. Friendly forces.

c. Weather (ceiling, visibility, wind, temperature, pressure and density altitude, sunrise and sunset, moonrise and moonset, percent of moon illumination, end evening nautical twilight, beginning morning nautical twilight, PZ and LZ altitudes, and weather outlook).

2. Mission. Clear, concise statement of the task that is to be accomplished (who, what, and when, and, as appropriate, why and where).

3. Execution.

a. Ground tactical plan.

b. Fire support plan to include suppression of enemy air defenses.

c. Air defense artillery plans.

d. Engineer support plan.

e. Tactical air support.

f. Aviation unit tasks.

g. Staging plan (both primary and alternate PZs).

(1) Pickup zone location.

(2) Pickup zone time.

(3) Pickup zone security.

(4) Flight route to PZ.

(5) Pickup zone marking and control.

(6) Landing formation and direction.

(7) Attack and air reconnaissance helicopter linkup with lift elements.

(8) Troop and equipment load.

h. Air movement plan.

(1) Primary and alternate flight routes (SPs, ACPS, and RPs).

(2) Penetration points.

(3) Flight formations) and airspeeds.

(4) Deception measures.

(5) Air reconnaissance and attack helicopter missions.

(6) Abort criteria.

(7) Air movement table.

i. Landing plan (both primary and alternate LZs).

(1) Landing zone location.

(2) Landing zone time.

(3) Landing formation and direction.

(4) Landing zone marking and control.

(5) Air reconnaissance and attack helicopter missions.

(6) Abort criteria.

j. Laager plan (both primary and alternate laager sites).

(1) Laager location.

(2) Laager type (air or ground, shut down or running).

(3) Laager time.

(4) Laager security plan.

(5) Call forward procedure.

k. Extraction plan (both primary and alternate PZs).

(1) Pickup location.

(2) Pickup time.

(3) Air reconnaissance and attack helicopter missions.

(4) Supporting plans.

l. Return air movement plan.

(1) Primary and alternate flight routes (SPs, ACPS, and RPs).

(2) Penetration points.

(3) Flight formations and airspeed.

(4) Air reconnaissance and attack helicopter missions.

(5) Landing zone locations.

(6) Landing zone landing formation and direction.

(7) Landing zone marking and control.

m. Coordinating instructions.

(1) Mission abort.

(2) Downed aircraft procedures.

(3) Vertical helicopter instrument flight recovery procedures.

(4) Weather decision by one-hour increments and weather abort time.

(5) Passenger briefing.

4. Service Support.

a. FARP locations (primary and alternate).

b. Ammunition and fuel requirements.

c. Backup aircraft.

d. Aircraft special equipment requirements, such as cargo hooks and command consoles with headsets.

e. Health service support.

5. Command Signal.

a. Signal.

(1) Radio nets, frequencies, and call signs.

(2) Communications-electronics operation instructions in effect and time of change.

(3) Challenge and password.

(4) Authentication table in effect.

(5) Visual signals.

(6) Navigational aids (frequencies, locations, and operational times).

(7) Identification friend or foe (radar) codes.

(8) Code words for PZ secure, hot, and clean; abort missions; go to alternatePZ and LZ; fire preparation; request extraction; and use alternate route.

b. Command.

(1) Location of air assault task force commander.

(2) Point where air reconnaissance and attack helicopters come under OPCONas aerial maneuver elements.

6. Time Hack. All watches are synchronized.

(Classification)

CHAPTER 4

Combat Operations

Section I

OFFENSE

4-1. General.

The air assault attack is the basic type of offensive operation conducted by an AATF. It is the integration of the combat, CS, and CSS elements in the movement into or out of an objective area. Generally, the term "insertion" applies when discussing the air assault into the objective area and the term "extraction" applies when discussing the air assault from the objective area. While these terms are fundamental to all air assault operations, they take on added importance in the attack.

4-2. Attack.

The opportunity to attack may arise during the course of battle, or it may be created by skillful, tactical leadership. Whatever the source, the attack is fast, violent, resolute, shrewd, and coordinated. There are two general types of attack: hasty and deliberate. The major difference between the two is time and enemy information available. The AATF may conduct an attack in conjunction with other forces. The type of action conducted by the larger force usually dictates the type of attack employed by the air assault force.

4-3. Hasty attack.

a. Situations in which an AATF might be called on to execute a hasty attack in support of a larger force are:

(1) During movement to contact by the larger force when unexpected contact is made. The AATF is committed to exploit a tactical advantage or to further develop the tactical situation.

(2) When part of the larger force's deliberate attack plan is modified while the operation is under way. The AATF is committed to reinforce in a weakened area or to exploit a tactical advantage.

(3) At the conclusion of an attack when a further advance is ordered. The AATF is committed to exploit the attack's success and to maintain momentum.

(4) An attack from a defense in which the commander sees an opportunity for offensive action and seizes the initiative.

b. When a hasty attack is considered under any of these circumstances, tentative PZs, LZs, and flight routes throughout the higher unit's zone of action are identified. This permits rapid commitment of the AATF anywhere in the sector. Because the hasty attack is conducted on short notice, there is little time to plan and orders are brief. The AATFC must rely on previous training and SOPs to cover these situations.

c. When the AATF is committed, the AATFC initiates several actions simultaneously. He directs suppressive fires to neutralize the enemy's ability to counter the air assault operation, and he concentrates sufficient combat power to overwhelm the enemy at selected points. While the AATF is en route, support fires suppress or destroy known or suspected enemy positions with priority of fires to the suppression of enemy air defense.

d. As the attack starts, attack helicopter units overmatch and react as necessary while the AATFC and FSO direct FA, mortar, CAS, and other supporting fires. FA and mortars destroy, neutralize, or suppress enemy indirect fire weapons as soon as they are located. Smoke may be used to screen aircraft movement from observation. However, the AATFC is careful that smoke does not obscure the LZ and hinder the landing operation. Airspace coordination must be effected early.

4-4. Deliberate attack.

a. The AATF, as part of a larger force operation, may conduct a deliberate attack. The AATF is provided sufficient time to develop a detailed, coordinated plan; receive additional assets; change task organization as necessary; and gather detailed intelligence. Detailed information about the terrain is collected so that best PZs, LZs, and flight routes can be selected. Air assault objectives are normally in the enemy's rear area, or the attack is from the flank or rear. This will normally preclude or limit the opportunity for leaders to see the terrain and force planners to rely on maps and aerial photographs.

b. When the larger force concentrates its combat power on a narrow front to break through enemy defenses, the AATF may bypass main defenses to destroy artillery positions, command posts, logistics and communications facilities, and/or to secure key terrain in the enemy's rear (Figure 4-1).

Figure 4-1. Landing away from the objective.

c. An attack against a heavier or well-prepared enemy force, particularly on the mechanized and/or armor battlefield, may subject the AATF to devastating firepower. For this reason, the AATFC may land the AATF away from the objective and conduct a dismounted attack in conjunction with friendly mechanized and/or armor forces. The AATFC must also consider that a highly mobile enemy force could encircle the AATF before it moves from an LZ. Consequently, he selects LZs in armor-restrictive terrain and employs antitank weapons and attack helicopter units against likely armor approaches. When used with accurate intelligence, these actions provide time to organize after landing and to attack the objectives.

4-5. Exploitation.

Exploitation is an operation undertaken to follow up success in the attack. Attacks are conducted with two overriding requirements: speed and violence. The attackers bypass pockets of resistance to concentrate on the destruction of the more vulnerable headquarters, combat support, and combat service support units. They disrupt the enemy's command and control; his flow of fuel, ammunition, repair parts; and his air defenses and artillery. This weakens and/or destroys the enemy. Enemy air defenses are avoided or suppressed for the AATF to exploit the situation.

4-6. Pursuit.

a. Pursuitisanoffensiveactionagainstretreatingenemy.Itspurposeistoenvelop the retreating force and destroy it by coordinated fire and maneuver. An AATF, operating as part of the pursuit force, can expect to be ordered to bypass resistance of any kind and move relentlessly to deep objectives that serve as checkpoints for the retreating enemy. The helicopter provides the AATF with the high degree of mobility required to conduct pursuit operations.

b. Air Force tactical aircraft, attack helicopters, and air assault forces can repeatedly attack the flanks of the withdrawing enemy columns, slowing them and aiding in their destruction. Blocking positions can be established on withdrawal routes to trap enemy forces between the encircling force and the direct-pressure force. Field artillery and forward arming and refueling points should be lifted into the encircling force areas as soon as possible.

4-7. Secure and defend.

a. This type of air assault operation is two-phased and requires detailed planning like a deliberate attack. The secure-and-defend mission is conducted when an objective, such as a vital terrain feature, must be seized and retained. The limited staying power of the AATF dictates early linkup with ground units, reinforcement by other units, or extraction from the enemy area.

b. The first phase is an attack to secure terrain to be controlled by the AATF in the initial stages of the assault. This should be a single-lift insertion of sufficient combat power to defeat enemy forces on the objective.

c. The second phase of the operation is the defense of the objective. The AATF normally establishes an airhead. This is a perimeter defense that controls all terrain essential to the defense of the objective. The airhead is large enough to provide operating space for combat, CS, and CSS units. It should include adequate LZs for simultaneous combat assaults using all airlift assets and provide space for landing artillery, follow-on forces, and supplies. The airhead is small enough for a battalion to defend yet large enough to permit defense-in-depth and maneuver of reserves to counter enemy attacks. As a rule, the area an infantry battalion can defend is 3 to 5 kilometers in diameter. Size is dictated by mission, enemy strength and disposition, terrain, and AATF combat power.

d. Boundaries delineate responsibilities of AATF subordinate elements. The airhead is often divided into company-size objectives for the air assault. Each company clears, secures and defends an assigned area of the airhead (which seldom has a rear area). The size sector assigned each company should be within its capability to seize and defend, based on an analysis of METT-T. Boundaries minimize adjustments during the transition from assault to defense. They should also prevent one unit from defending in widely divergent directions. A company facing a dangerous avenue of approach, for example, is assigned a smaller sector than a company facing a less dangerous avenue. Defensive responsibility for an avenue of approach is not divided. The unit assigned the approach also covers any dominating terrain.

e. A terrain feature to be secured in the assault, and vital to AATF mission accomplishment, is designated an assault objective. The assault objective should include terrain that dominates all high-speed approaches into the airhead area. Assault objectives are assigned priorities. Those specified by higher headquarters are given first priority. Others are ranked according to the threat they would pose if controlled by the enemy. A company's sector should include at least one LZ for the assault and to aid in resupply and evacuation.

4-8. Reconnaissance in force.

a. A reconnaissance in force is conducted to determine or test the enemy's disposition and strength or to develop intelligence. It is conducted when the enemy situation is vague. This type of operation is conducted by forces strong enough to accept engagements with the enemy in order to accomplish their mission. The information obtained (for example, major weaknesses in enemy positions), if promptly exploited, may provide a significant tactical advantage. The reconnaissance in force is planned and conducted with elements specifically prepared to find the enemy and develop the situation. Once the units are committed, they are on a "be prepared to fight" status.

b. The reconnaissance in force is an ideal mission for the AATF in an insurgent environment in order to keep constant pressure on a guerrilla force. The AATF is suited for reconnaissance-in-force operations against conventional light infantry forces. However, it is not suited for such operations in a strong armor threat area due to the likelihood of ground contact with an enemy force that has superior firepower, mobility, and protection.

c. The reconnaissance in force accepts risk to gain intelligence information rapidly and in more detail than other reconnaissance methods. The commander assigning an AATF this mission must determine the following:

(1) Isthedesiredinformationimportantenoughtojustifytheriskstopersonnel and equipment?

(2) Can other intelligence methods obtain the same information in sufficient time with less risk?

(3) Will the reconnaissance in force compromise future plans?

(4) Can the operation succeed?

d. The reconnaissance in force, however, differs from the normal attack that is conducted to destroy enemy forces or secure terrain. The reconnaissance in force locates the enemy and presses him into reacting. When the force discovers a weak point, the AATF exploits it quickly. The AATFC exercises caution, however, since the enemy response may be too strong for the AATF. Thus, the commander also plans withdrawal to avoid destruction of the AATF.

e. When the commander wants information about a particular area, the reconnaissance in force is planned and executed as an attack against a specific objective. The objective is of such importance that, when threatened, the enemy will react. For example: a successful reconnaissance in force may cause commitment of enemy reserves, redeployment of enemy fire support means, or adjustment of enemy second-echelon forces. Taking a terrain objective is not in itself the purpose of the operation. Rather, the operation seeks to obtain specific information about the enemy by seizing a terrain objective. The objective location depends on the information desired. The AATF's combat power must be sufficient to force enemy reaction. This should disclose positions, strength, planned fires, and planned use of reserves. It may also disrupt the enemy's planned operations and take the initiative from him. An AATFC can deploy all three companies against specific objectives; or the commander may commit one or two companies and retain the third to respond to tactical situations as they develop. When the enemy reacts to one unit, the units not in contact are shifted to exploit revealed enemy weaknesses or help extract a unit under pressure.

4-9. Raid.

a. **Air assault raid.** This is a swift penetration of hostile territory that may be conducted to destroy installations, confuse the enemy, or gather information. It ends with a planned withdrawal. Because a raid is conducted behind enemy lines, it requires exact planning to ensure a high probability of success. The selection of LZs, PZs, and flight routes (as in the deliberate attack) is based on the results of detailed planning and ample intelligence. Since the raiding force attempts to achieve surprise, the decision to land on the objective takes on added significance (Figure 4-2).

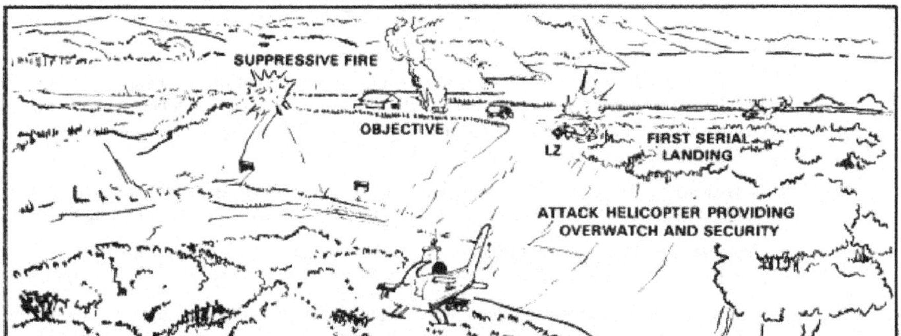

Figure 4-2. Operation of a raid.

(1) The AATF may land on or near the objective when the following applies:

(a) There is a suitable LZ.

(b) The enemy does not have a highly mobile reaction force nearby to attack the AATF immediately after it lands.

(c) The objective is not accessible overland.

(d) The AATFC determines that overland movement would expose his forces to enemy fire and possibly disrupt the mission.

(e) Surprise is important.

(f) When there are no armor or mechanized units and/or vehicles on the objective.

(g) When there are no air defense weapons on the objective.

(h) When the AATF can land overwhelming combat power quickly on the objective.

(2) The AATF should land some distance from the objective when these considerations apply:

(a) The AATFC decides to assemble and reorganize before conducting the assault.

(b) The only suitable LZs are away from the objective.

(c) There is a highly mobile enemy force on the objective that could disrupt the landing.

(d) Surprise is not imperative.

(e) Local air defense is too strong.

b. **Conduct a raid.** The AATFC task organizes his force to accomplish four essential tasks:

(1) Command and control.

(2) Security.

(3) Support.

(4) Assault.

(a) **Command and control.** The AATFC commands and controls from a location that offers the best vantage points; however, during air movement, the AATFC normally uses a command and control helicopter. After units are on the ground, he may join one of the ground units or he may continue to control from the air.

(b) **Security.** The element given the security mission blocks avenues of approach into the objective and provides suppressive fires for withdrawal after the mission is complete.

(c) **Support.** The element providing assault support lays down a heavy volume of suppressive fire to neutralize the objective and destroy the enemy that is occupying it.

(d) **Assault.** The element conducting the assault secures the objective and provides security for specialty teams (for example, demolitions).

c. **Rehearsal.** Rehearsals are critical to success. The operation should be rehearsed several times by all elements participating in the raid. If it is a night raid, rehearsals are conducted during daylight and darkness.

d. **Withdrawal.**

(1) A raid differs from other attacks in that it includes a withdrawal plan. The plan contains provisions for withdrawal by air as well as on foot (in case aircraft cannot extract the force). Withdrawal on foot may require the entire force to move as a unit, or the force may have to break down into small elements to evade enemy contact and exfiltrate the area.

(2) A withdrawal by air involves movement into, and defense of, the PZ (Figure 4-3). If the withdrawal is for the purpose of committing the force to another combat mission, then an additional ground tactical plan is prepared for that phase of operation. Either plan is as detailed as time permits and includes:

(a) Pickup zone designation.

(b) Fire support plan for movement to secure the PZ and to cover the withdrawal.

(c) Schedule of unit movement to the PZ.

(d) Loading priorities.

(e) Designation of, and instructions for, the PZ control group.

(f) Landing zone designation following withdrawal.

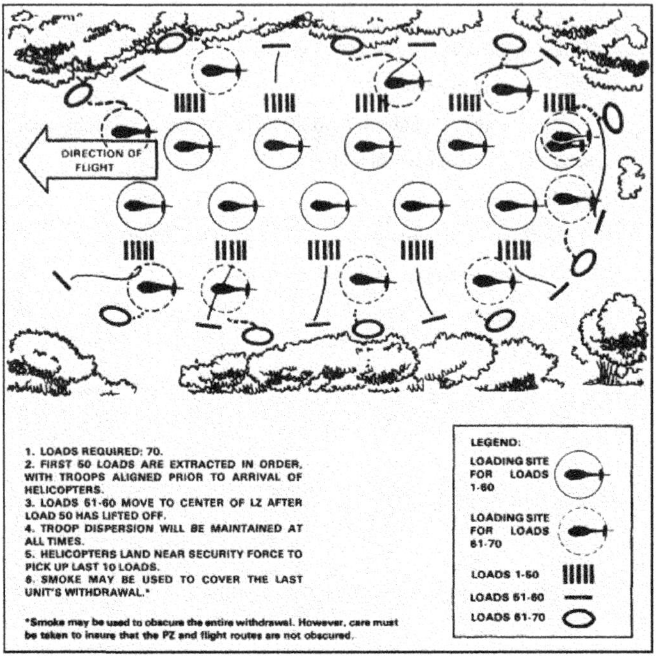
DIRECTION OF FLIGHT
1. LOADS REQUIRED: 70.
2. FIRST 50 LOADS ARE EXTRACTED IN ORDER, WITH TROOPS ALIGNED PRIOR TO ARRIVAL OF HELICOPTERS.
3. LOADS 51-60 MOVE TO CENTER OF LZ AFTER LOAD 50 HAS LIFTED OFF.
4. TROOP DISPERSION WILL BE MAINTAINED AT ALL TIMES.
5. HELICOPTERS LAND NEAR SECURITY FORCE TO PICK UP LAST 10 LOADS.
6. SMOKE MAY BE USED TO COVER THE LAST UNIT'S WITHDRAWAL.*
*Smoke may be used to obscure the entire withdrawal. However, care must be taken to insure that the PZ and flight routes are not obscured.
LEGEND:
LOADING SITE FOR LOADS 1-60
LOADING SITE FOR LOADS 61-70
LOADS 1-50
LOADS 51-60
LOADS 61-70

e. **Unit sequence.** The sequence of unit withdrawal varies according to the tactical situation and the subsequent mission of the AATF. Administrative, combat service support personnel, and heavy equipment are withdrawn first (before tactical units). The commander may withdraw them to a secured (intermediate) area and then move them to another (combat) area after it is secured. Unit redeployment is determined by combat and security requirements in the new area.

f. **Pickup zone designation.**

(1) **Pickup zones.** They are designated by the headquarters controlling the withdrawal. PZs are as close to unit positions as the terrain and enemy situation permit. To achieve speed in landing, loading, and lift-off, multiple PZs may be used (consistent with available security forces).

(2) **Pickup zone control officer.** A PZCO is designated for each unit's PZ. He is responsible for calling units and guiding them from their assembly areas to the PZs to expedite loading. The senior PZCO coordinates all PZ operations when using multiple PZs (and PZCOS). He maintains contact with the AMC to ensure coordinated arrival of troops and aircraft.

(3) **Security.**

(a) Security elements are positioned to cover the main body as it assembles, moves to the PZ, and withdraws. Security may be composed of small detachments from each subordinate unit, or it may be one of the subordinate units. The latter is usually best. Unit integrity aids control and gives more effective reaction in case of attack. Security elements protect the PZ at a time ordered by the unit commander conducting the withdrawal.

(b) Each aircraft withdrawing the security force lands as close as possible to its individual load. During this short interval, attack helicopter teams overmatching the withdrawal provide security. Panels, or other covert markers, identify each loading site (when the withdrawal is conducted during limited visibility, chemical lights or directional beacons are used).

(c) The last security element to withdraw achieves some protection by detonating Claymore mines and firing automatic weapons just before loading. The loading and lift-off are executed quickly. Attack helicopter teams overmatch the lift-off.

(4) **Reserve.** A reserve, when designated, may remain airborne near the PZ or on standby in another area. This gives the commander a reaction force that can be employed as required to support withdrawal of the security force.

(5) **Fire support.** During withdrawal, fire support is planned and executed to protect security elements as combat power on the PZ diminishes.

Section II

DEFENSE

4-10. General.

Defense is a coordinated effort by a force to defeat an attacker and prevent him from achieving his objectives (FM 100-5). Army doctrine does not describe the form of defense to be used in battle because there is literally an infinite number of defensive techniques that a commander can select from in developing a defense. The commander, by using the METT-T analysis, can determine the best technique for a specific tactical situation.

4-11. Defensive operations against an infantry-heavy threat.

a. **Air assault task force defending.** The AATF defends against an infantry-heavy Threat by employing its airmobility to achieve a maneuver advantage over the enemy. This advantage allows it to perform operations in the covering force area (CFA), main battle area (MBA), and rear area.

b. **Covering force area.** The AATF can conduct covering force operations for a larger force. Normally, the covering force consists of air reconnaissance, infantry, artillery, engineer, and attack helicopter units. Infantry and artillery assigned to the covering force must be provided enough helicopters to move the entire unit. The covering force is generally organized based on the:

(1) Number of enemy avenues of approach into the CFA.

(2) Size and type of enemy forces.

The covering force accomplishes its mission by placing the majority of its combat units on the most dangerous avenues of approach into the CFA. Air reconnaissance deploys to the front and provides early warning of the direction, speed, and composition of enemy forces. Enemy units are taken under fire as soon as they are within range of weapons. As the enemy attempts to close with ground units of the covering force, attack helicopters, artillery, and close air support provide firepower to enable ground units to displace by air to successive positions. Protection of air assault infantry and antitank systems is achieved by superior mobility. Covering force units attrite the enemy, deceive him as to the location of the MBA, slow his speed of advance, cause him to mass, and may cause him to divulge his intentions. Units are assigned subsequent missions in the MBA when the covering force mission has been accomplished.

c. **Main battle area.** The mobility advantage which the AATF has over enemy infantry-heavy units allows it to defend in greater depth. The AATF defends by orienting on the destruction of advancing enemy forces and fights a series of battles in depth, attacking the enemy from the front, flanks, and rear while using minimal forces to maintain surveillance over the remainder of the assigned sector. Battle positions are selected and prepared throughout the MBA along likely avenues of approach. Primary and alternate LZs and PZs are selected for each battle position. When enemy fires preclude extraction of the AATF from battle positions, covered and concealed routes are selected for foot movement to alternate PZs. Only when absolutely necessary should an AATF be directed to occupy or retain terrain.

If there is a situation in which the retention of terrain is essential to the defense of the entire sector, its retention is specified.

4-12. Defensive operations against an armor-heavy threat.

a. **Air assault task force defending.** An AATF is not well-suited to perform a defensive operation against mounted forces on terrain favorable to mounted operations. However, the AATF can effectively operate in the restrictive terrain of built-up areas or mountains or defend chokepoints,

b. **Types of operations.** The AATF can conduct the following operations on the armored and mechanized battlefield in support of larger defensive operations:

(1) Main battle area operations in restrictive terrain.

(2) Economy of force or reserve.

(3) Rear operations.

(4) Flank security operations.

(5) Limited-objective counterattack operations or raids.

(6) Delay and withdrawal operations.

(7) Seizure of FLOT objectives for linkup operations.

c. **Covering force area.** Attack helicopter and air reconnaissance units are the elements best suited for employment in covering force operations when employed with armored and mechanized units.

d. **Main battle area.** The AATF is not well-suited to defend against armored and mechanized forces. If it is used to defend against such forces, it should be employed in restrictive terrain not favorable to employment of massed armor. The AATF can be employed in built-up areas, mountainous terrain, and heavily forested areas. Attack helicopters can be employed as a mobile, tank-killing reserve.

4-13. Economy of force.

Defense in an economy-of-force role can be accomplished by displacing units of the AATF in depth on the avenue of approach throughout the sector. The air reconnaissance elements can screen areas where enemy attack is possible but unlikely. Combat units are repositioned to counter the major enemy thrust. After engaging the enemy, and before the enemy closes on battle positions, units are picked up from designated PZs and organized in depth. The AATF essentially conducts a delay. Field artillery is repositioned as necessary to halt the enemy advance. Attack helicopter elements should be placed under the operational control of the AATFC. Elements of the AATF held in reserve are rapidly transported by helicopter into areas under enemy pressure.

4-14. Delay.

a. The key to success in the delay is the commander's ability to array forces in depth before the initiation of the delay. Decisive engagement is accepted only to the degree and extent necessary to accomplish the delay mission. Contingency plans for stay-behind operations should be developed. The AATFC should continually look for and seize the opportunity to launch small-scale offensive air assault and attack helicopter raids into the enemy flanks and rear areas. A delay may be conducted to:

(1) Gain time so that other forces can deploy.

(2) Serve as an economy-of-force measure to allow concentration of friendly forces in other areas.

(3) Determine enemy composition, strength, intentions, and capabilities.

(4) Channel the enemy into selected areas and then destroy him.

b. The AATF is seldom given a "timed-delay" mission. This type of mission would require an AATF to delay the enemy for a specified time and would restrict its mobility and subject it to unacceptable losses.

c. The delay-in-sector mission is more appropriate. The TF disengages by helicopter before it is decisively engaged. Against armor forces, the AATF should displace at distances of no less than 1,500 meters and rely on attack helicopters to delay the armor while friendly infantry is extracted.

Section III

OTHER TACTICAL MISSIONS

4-15. General.

This section discusses eight types of operations. Any one may be applicable in an air assault mission.

4-16. Screening.

a. An AATF screening force provides early warning over an extended frontage. Screening missions are assigned to:

(1) Provide timely warning of enemy approach.

(2) Maintain visual contact and report on movement.

(3) Destroy or repel small enemy forces by employing organic and/or supporting fires.

(4) Impede the advance of larger forces and destroy the enemy by employing long-range organic fires.

b. A screening mission employs a series of observation posts (OP) overlooking enemy avenues of approach and the areas between them. Patrols cover dead space between OPs and cover other areas during limited visibility. When contact is established, the screening force withdraws on order, maintaining visual and/or electronic contact, and reports enemy movements. As in the delay, timely displacement is critical to AATF survival.

4-17. Guard force.

The AATF can perform flank or rear guard missions for a division or larger force and help protect the main body from ground observation, direct fire, and surprise attack. As a guard force, it has sufficient combat power to attack enemy reconnaissance forces and to delay an enemy attack until the main body can deploy. The rear guard follows the main body, occupying successive positions. The rear guard also screens between flank positions and rear elements of the main body. The AATF conducts rear guard operations by moving from position to position. These movements are controlled by using designated phase lines.

4-18. Covering force.

The air reconnaissance squadron of the division can overfly rough terrain, find the enemy, and develop the situation. Brigades are deployed as necessary to ensure the uninterrupted movement of the main body. The brigade may use one of the following two methods to conduct the division covering force mission:

a. Reinforced air reconnaissance elements under divisional control reconnoiter while the air assault task force remains in assembly areas or on order to be available for commitment. When contact is made with the enemy, and after the air reconnaissance has developed the situation, AATFs are committed to destroy the enemy.

b. Brigades, with air cavalry elements under their operational control, conduct covering force operations as the division minus moves by bounds behind the leading brigade.

4-19. Reinforcement of committed units.

a. An AATF can reinforce a committed unit in three ways:

(1) With uncommitted units (reserves).

(2) With additional antitank (AD platoons.

(3) By moving field artillery to weight the battle.

b. Brigade commanders may direct the insertion of an AATF unit to reinforce threatened sectors or add depth to the battle area.

c. Antitank platoons may be taken from a reserve unit or a unit that is not protecting an armor approach. Depending on the number of sections employed, the AT platoon leader and/or the platoon sergeant accompanies them for command and control. Tube-launched, optically tracked, wire-guided (TOW) missile sections are used for their long-range fires and accuracy. Careful consideration is given to planning the extraction of AT platoons because they lack ground mobility. An infantry squad may accompany each section to provide security. The unit receiving an AT section assumes responsibility for effective positioning, integrating their fires with other elements, and providing necessary support.

d. Firing batteries can be rapidly shifted about the air assault battlefield to ensure fire support to committed units.

4-20. Linkup operations.

a. When withdrawal of an AATF from the objective area is not planned or feasible, a linkup operation is conducted to join two forces. The AATF may participate as part of a larger force, or it may conduct a linkup with its own resources. Close coordination and detailed planning between the commanders of both units are essential. Some of the things that must be coordinated are:

(1) **Command relationships.** To delineate responsibilities, it is necessary to specify who assumes command upon linkup. The headquarters directing the linkup determines the command relationship, its effective time, and the responsibilities of each force during the operation.

(2) **Liaison and responsibilities.** Once command relationships are established, the commanders of the units involved establish liaison. If conditions permit, the commander and liaison teams meet face to face; if not, then coordination is accomplished by radio or other available means, such as messengers. During the operation, the units attempt to maintain continuous radio contact with each other or with higher headquarters. As a minimum, the units exchange the following information:

(a) Enemy and friendly situations.

(b) Locations and types of obstacles (natural and artificial).

(c) Fire support plan.

(d) Air defense control measures.

(e) Recognition signals.

(3) **Mutual recognition signal system.** A system of mutual recognition signals is established and made known to all units participating in the linkup. Provisions are made for recognition procedures that may be used both day and night for air and ground elements alike.

(4) **Communications.** It is the responsibility of the headquarters directing the linkup to ensure communications-electronics operation instructions compatibility. If the linking units do not have the same CEOI, the higher headquarters directs one unit to change. The unit to change is normally the one not in contact, encircled, or breaking out. This is especially critical for recognition signals. If the units involved in the operation are neither under OPCON nor attached, they maintain their parent command nets.

(5) **Schemes of maneuver.** Both units' schemes of maneuver, to include control measures, are exchanged. When a passage of lines is required after linkup, the control measures include primary and alternate linkup points, start points, routes, and release points.

(6) **Fire support.** Fire support coordination measures are established and disseminated to both forces by the headquarters ordering the linkup. A restrictive fire line (RFL) is normally established on identifiable terrain as close as possible to the stationary force. The RFL prohibits the fires and the effects of fires from extending across the line without coordination with the affected force. Each force must be prepared to support the other as the situation dictates. Upon linkup, or at some prearranged time, control of supporting fire becomes the responsibility of the commander previously designated to have overall responsibility for the operation. Since the maneuvering unit is normally larger and has more supporting weapons and more reliable supply

lines than the stationary AATF, the maneuvering unit can provide more support than the stationary unit. However, the stationary unit, since it is already in position, can provide some support to the maneuvering unit. Such support is limited by the amount of ammunition and the number and type of weapons assigned to the stationary unit.

(7) **Actions after linkup.** These are specified in the order given to the units conducting linkup. This ensures operational continuity and reduces massing of units. Actions may include reinforcing the defense of the area, conducting a coordinated attack, or passing the maneuver unit through the stationary unit to continue the attack.

(8) **Assistance.** This includes the mutual assistance that the stationary and maneuvering units can provide to each other. Because of its lack of assets in the airhead, the stationary unit normally can provide only limited assistance to the maneuvering unit. The stationary unit, however, can normally provide the following:

(a) Guides.

(b) Lanes through obstacles and the airhead.

(c) Traffic control.

(d) Limited logistical and maintenance support.

(e) Limited medical support (for example: holding areas for, and possible evacuation of, dead and wounded).

(f) Limited fire support.

(g) Information on recent enemy activity.

The maneuvering unit normally provides more assistance because of its established lines of communications. However, if it experienced heavy combat during the move to the linkup, this assistance may be reduced. The maneuvering unit can normally provide logistical, maintenance, medical, and fire support.

(9) **Alternate plan.** An alternate plan is developed to cope with unexpected enemy activity.

b. Once the airhead is established, the linkup points are occupied. When the maneuvering unit is within range and identified, communications are established between it and the stationary unit. Once linkup is effected, the two units follow the procedures for a passage of lines, and they continue their assigned missions.

4-21. River crossing operations.

AATFs by their makeup, may reduce CSS considerations during river crossing operations. Such forces may overfly the river or support bridge construction. Reconnaissance elements can be deployed by air to verify and collect essential intelligence on crossing sites and enemy dispositions. Objectives can be reached on the far shore quickly, eliminating enemy interference with development and use of crossing sites. Engineer bridging assets can be airlifted forward rapidly, eliminating traffic problems on the crossing site approaches. If a deliberate crossing is chosen, the AATF, with its increased mobility, can be used to clear the near shore of enemy resistance. During the actual crossing, whether it be hasty or deliberate, the AATF can assist by:

a. Attacking enemy forces that interfere with the crossing by seizing objectives thatwould secure, or assist in securing, the bridgehead.

b. Providing flank security.

c. Securing crossing sites.

d. Screening the crossing sites with smoke.

4-22. Rear operations.

a. Countering enemy airmobile, airborne, or guerrilla infiltration threats, the AATF monitors likely infiltration routes and probable target areas for airborne or airmobile attacks. Probable LZs and PZs are identified and monitored by observation posts or remote sensors. Potential infiltration routes in unoccupied terrain are monitored with sensors to detect the enemy as early as possible.

b. Air reconnaissance units provide wide-area surveillance and security, and are integrated into reaction force plans.

c. Rear operations are coordinated with designated military police, civil affairs groups, and other civil and military organizations. The AATF, as the initial reaction force, contains the enemy force if it does not have enough combat powers to destroy it, and relies on additional forces to destroy the enemy.

4-23. Limited visibility operations.

a. A commander may desire to take advantage of limited visibility conditions to gain maximum surprise or deception, maintain the momentum of successful operations, reinforce or withdraw committed units, and/or deploy maneuver support elements.

b. The following aircraft operational requirements must be considered:

(1) Desired directions) and route(s) of movement for aircraft (to include identification of selected terrain feature).

(2) The identity and location of LZs and/or PZs.

(3) Emergency ground-to-air signals.

(4) Directions and points of landing for aircraft.

(5) The presence of LZ obstacles is indicated to the aircraft flight commanders through electronic and/or visual navigation aids.

c. Some advantages of limited visibility operations are:

(1) Aircraft are partially concealed from enemy visual observation.

(2) Maximum surprise and confusion can be achieved.

(3) Continuous pressure can be exerted on the enemy.

(4) Effective enemy air defense fire, and interdiction by enemy aircraft, are diminished.

d. Disadvantages of limited visibility operations also exist. The need for more elaborate control measures and caution on the part of the aviators and troops slow operations. However, with proper equipment, constant training, and a thorough knowledge of techniques, these disadvantages may be overcome. The following factors are considered:

(1) More time is required for planning, preparation, and execution.

(2) Formation flight is more difficult, and formations are more dispersed.

(3) LZs and/or PZs used should be larger.

(4) Navigation is more difficult.

(5) Additional illumination is planned and immediately available to the AATFC in case it is necessary for mission accomplishment.

4-24. Operations in a nuclear, biological, and chemical environment.

a. In the event of a nuclear attack, AATFs can conduct a radiological survey and, when feasible, move into the target area after the explosion to stall enemy exploitation of its effect. AATFs can rapidly and safely bypass obstacles created by a nuclear strike, whether their objective is within or beyond the target area.

b. When planning air assault operations in conjunction with friendly nuclear munitions employment, the planner must consider:

(1) Effects of intense light on pilot vision.

(2) Effects of intense heat on equipment and personnel.

(3) Effects of blast waves on aircraft in flight.

(4) Residual radiation rates on the LZs.

(5) Utilization of LZs; debris may prohibit their use.

(6) Effects of electromagnetic pulse (EMP) on electronic equipment.

(7) Selection of approach and departure routes into contaminated LZs.

(8) Use of alternate LZs when primary LZs are judged as having too high a residual radiation rate.

c. Planning for air assault operations in a toxic environment includes consideration of the following:

(1) Reconnaissance of areas known or suspected of contamination.

(2) Selection of routes and positions with regard to contaminated areas to avoid stirring up or spreading agents with rotor wash.

(3) Protection of supplies and equipment.

d. The three principles of NBC operations (contamination avoidance, protection, and decontamination) are fundamentals that ensure survival (see FMs 3-3, 3-4, and 3-5).

e. If air assault operations must be conducted following contamination, the AATFC may direct that hasty (spot) decontamination of aircraft be accomplished. Spot decontamination is an effective means of decontaminating specific areas of an aircraft. This sustains flight operation since certain functional areas are treated before they are touched. Surfaces must be washed with decontaminants to flush agents off the aircraft skin. Small amounts of the NBC agent (absorbed into the fuselage paint) will probably remain after decontaminating. The evaporation of these residues can create a vapor hazard; therefore, personnel in and around the aircraft continue to wear the protective mask and gloves. Decontamination reduces the hazard of agent contact and transfer. Six functional areas applicable to spot decontamination are:

(1) Refueling procedures.

(2) Rearming procedures.

(3) Entry and exit from the aircraft.

(4) Preflight and postflight inspections.

(5) Maintenance inspections.

(6) Battle damage repair.

Aviation personnel are trained in spot-decontamination procedures but may require equipment to effect all required decontamination quickly.

CHAPTER 5

Combat Support

Section I

COMBINED ARMS TEAM

5- 1. General.

The air assault task force commander uses combat support elements to enhance the combat power of his maneuver elements. Knowing combat support capabilities, assigning them appropriate missions, and controlling their operations are essential to the application of superior combat power at the decisive time and place. The AATFC's key role in integrating combat support elements with his maneuver elements, to form the combined arms team, is critical for success in the AirLand Battle.

5-2. Employment of combat support.

a. **Combat support elements.** They are normally under OPCON or in DS of the AATF in order to ensure the close coordination and continuous, dedicated support required in air assault operations.

b. **Determining task organization.** The AATFC assigns an element that is attached, under OPCON, or in DS, to one of his subordinate maneuver units when he feels the element could be more effectively controlled or employed by one particular unit, rather than under AATF control. General support is used when the combat support element can best support the operation under centralized control to quickly shift its efforts to the point needed (for example, mortars that are normally used in a general;support role) and when the situation is vague or changing.

c. **Basic responsibility to support,** No matter what support status the combat support elements are assigned, the AATFC has the responsibility to ensure the combat support units are properly supported by the AATF. Although the AATFC is not required to provide support under the status of DS, GS, or OPCON, it is to his advantage to ensure the CS elements are properly supported. This means providing rations, fuel, and ammunition as required. It also means expediting repair of equipment outside the capabilities of the AATF maintenance unit, The advantage of doing this is to ensure the CS elements are able to continue providing support.

d. **Relationship to the AATF staff.** The commander of the combat support unit must be both a commander and a special staff officer. This means he commands his unit and provides advice and assistance to the AATFC. He serves as a special staff officer during the planning phase of an operation, providing assistance and advice in the preparation of the operation order. He can also provide advice and assistance during the conduct of the operation, but this is limited since his primary concern is command of his unit.

Section II

FIRE SUPPORT

5-3. General.

a. Fire support is the collective and coordinated employment of mortars, field artillery, attack helicopters, close air support, naval gunfire, and other fires in support of battle plans. The mission of the fire support system is to destroy, neutralize, or suppress surface targets in support of air assault operations. It includes suppression of enemy air defenses which is imperative for air assault operations.

b. The AATFC integrates the firepower of mortars, field artillery, close air support, EW, and, when available, naval gunfire, with the maneuver of combat units to defeat the enemy. Fire support enhances the AATF's combat power by:

(1) Destroying, suppressing, and neutralizing targets.

(2) Obscuring the vision of enemy forces.

(3) Isolating enemy formations and positions.

(4) Slowing and canalizing enemy movements.

(5) Killing or disabling the enemy at ranges greater than that of direct fire weapons.

(6) Screening with smoke or creating obstacle areas with the employment of scatterable mines.

(7) Reducing the effects of enemy artillery by active counterfire.

(8) Interdicting follow-on enemy echelons.

(9) Providing illumination.

c. To effectively utilize the fire support assets, the AATFC must have an understanding of the field artillery support relationship. The artillery force commander is the fire support coordinator for the AATF, and the fire support officer is the assistant FSCOORD. Each AATF is provided a fire support element, led by an FSO, from the direct support battalion. In those instances when the AATF is operating independently, it may be necessary to attach an artillery unit (battery or battalion) to provide adequate fire support. Attachment is a nonstandard mission and involves special considerations for the AATF, such as the responsibility to provide security, logistical support, and lift capability to the artillery unit.

d. Fire support planning techniques and measures specific to air assault operations are discussed in Chapters 3 and 4.

e. Appendix F includes a discussion of specialized training required for support of air assault operations.

5-4. The fire support coordinator.

While the AATFC is responsible for the integration of all fires with the maneuver plan, the FSCOORD is his principal assistant for the proper integration and application of fire support. Working together as a team, the supported commander and the FSCOORD generate the maximum combat power available to support the ground forces.

5-5. Fire support delivery systems.

The AATF is unique in its mission and organization and so are its support elements. They are specifically tailored to be integrated into the AATF. The indirect fire assets must be light and maneuverable and capable of maintaining the fast pace of the AATF. The fire support delivery means available to the AATF may include:

a. **Mortars.** Organic to each infantry battalion, they are used to provide close-in direct fire support.

b. **Artillery.** Supporting artillery must either be positioned well forward to provide fires from the PZ to the objective area, or must be air lifted with the AATF to the objective area.

c. **Air defense artillery.** Supporting air defense artillery can be called upon to provide direct fire support when the situation demands and the commander so directs.

d. **Close air support.** In most cases, United States Air Force (USAF) aircraft will be available to provide close air support. Requests for these aircraft are processed through the TAC CP collocated with the AATF.

e. **Naval gunfire.** Navy cruisers and destroyers provide fire support in coastal areas. Naval gunfire spotters from a United States Marine Corps (USMC) air and naval gunfire liaison company (ANGLICO) may be attached to the AATF to control these fires.

f. **Attack helicopters.** Because of their mobility and firepower, attack helicopters may be integrated into the fire support plan when conventional field artillery is not available.

Section III

FIRE SUPPORT PLANNING AND COORDINATION

5-6. General.

a. Fire support planning addresses how fire support is to be used to support maneuver forces. Fire support coordination entails those actions needed to implement plans and manage resources on the battlefield. Although planning and coordination are separated, they overlap and are mutually supporting. If the planning has been done well, the implementation (coordination) will give the commander the support he needs to win. For a discussion of fire support planning and coordination, see Chapter 3, FM 6-20.

b. The planning and coordination process begins when the mission is received or assumed. The AATFC, S3, and the FSO interact throughout the planning sequence, the decision process, and the execution of the mission.

5-7. Facilities.

At the AATF level, the FSO advises the AATFC on how fire support can best influence the operation. He performs the planning and coordination of fire support assets to include mortars, field artillery, close air support, and naval gunfire. The fire support element (FSE) and fire support team (FIST) provide personnel for continuous planning and coordination of support fires.

a. AATF fire support element. The fire support element at each AATF (battalion-size) consists of an FSO, assistant FSO, fire support sergeant, and fire support specialists. The FSE, AATF S3 Air, and advisers from the other fire support means are collocated within the AATF TOC for the planning and coordination of fire support. The FSE coordinates and works closely with the brigade FSE, and FSEs of other battalions, the DS field artillery battalion fire direction center (FDC), and S2 and S3, the tactical air control party, aviation liaison officer, S3 Air, the mortar platoon leader, engineer platoon leader, and the company fire support teams. The FSO supervises the operations of the FISTS.

b. Company fire support team. The company fire support team and the forward observer (FO) section provide the fire support planning and coordination for maneuver companies and platoons. The FISTs are provided by the DS FA battalion. Occasionally, firepower control teams for naval gunfire (NGIN and naval air, and forward air controllers for USAF CAS collocate at the company FIST to advise and assist in the use of their assets. The FIST is supervised by an FA lieutenant who serves as the company commander's fire support officer.

5-8. Fire support planning.

The planning process determines how fire support will be used: what types of targets will be attacked, when, and with what means. It is sufficiently flexible to accommodate the unexpected in combat. Integrated fire support can result only when the FSCOORD is an aggressive contributor to the AATFC's planning sequence and decision making process.

a. The depth and detail of fire support planning depend on how much time is available. Many of the actions that occur in response to battle situations are established in SOPs and in fragmentary orders (FRAGO).

b. Fire support planning is continuous and concurrent at all levels. During the battle, planning is concurrent with fire support coordination to implement the fire support plan on the battlefield.

c. The fire support plan outlines the way artillery, mortars, close air support, and naval gunfire are used to complement the scheme of maneuver, and provides instructions for executing those fires. It also details the use of AATF target acquisition assets. It prioritizes targets, matches them with the available fire support systems, and allows fires to be executed quickly (without specific direction from the commander) once the operation starts. An AATF fire support plan should include:

(1) A detailed concept of how fires support the air assault operation from the initial PZ to the final objective.

(2) A target list that includes locations where fires are expected or likely to be used.

(3) A priority of fires telling which element receives fire in case of conflict (for example,

priority of FA fires to Team A; mortar fires to Team B).

(4) Target attack priorities establishing which type of mission to fire first in case of conflict (for example, first priority to enemy air defense systems; second priority to assist disengagements).

(5) An allocation of priority targets to indirect fire assets, if designated.

(6) Firing schedules for the indirect fire weapon systems. This planning tool identifies who fires the mission, when it occurs, and the nature of fires (for example, family of scatterable mines, smoke, SEAD preparations).

(7) Informal airspace coordination areas (ACA).

(8) Coordination measures for providing troop safety and promoting synchronization of supporting fires.

d. During the planning of fire support for an air assault operation, the FSO must consider displacement. When FA can support the AATF from a secure area (without displacement forward of the forward edge of the battle area [FEBA]), it does so. If such support is not feasible, the FSO determines if other fire support is sufficient to accomplish the mission. If other support is not sufficient, it may be necessary to displace the FA into the objective area. When the decision to displace is made, consider that:

(1) Displacement is accomplished by echelon to prevent temporary loss of FA support.

(2) Field artillery requires security in the objective area.

(3) Cargo helicopters are required to displace the FA unit.

(4) Ammunition resupply is made by air.

(5) The FA depends on helicopter assets for mobility unless prime movers are lifted into the objective area.

(6) Supporting, towed artillery (M198 or lighter) must be available.

e. The fire support plan is developed by the FSO with assistance and input from the Air Force LO, FISTS, heavy mortar platoon leader, S2, and S3. A fire plan is constantly refined or modified as the operation continues. Thus, the fire support plan facilitates responsive fires to the AATF wherever they are needed.

f. A formal and/or informal planning approach at the AATF level is a combined process that uses the principles of both formal (downward) and informal (upward) planning. Initially, the AATF FSO disseminates, in the operation order, a fire support plan to support the AATF. This product usually contains all the elements listed above. The FSO plan is modified as company and/or FIST fire plans are received. The rewritten fire plan is disseminated to each weapon system for execution.

g. To facilitate fire planning, company FSOs normally accompany team commanders to the AATF OPORD briefing. This permits the company FSOs to hear the operational concept simultaneously with their commander. Within minutes after the OPORD, they can get together to develop their fire support plans. This arrangement also allows the AATF FSO to brief the company FSOS on plans the AATFC wants implemented. Written fire plans can be disseminated. Questions can be answered quickly and conflicts can be resolved with minimum confusion.

h. Suppression of enemy air defense is a critical task in fire support to ensure success of the air assault operation and must be planned. Lift helicopters are especially vulnerable to enemy air defenses. Unless there are overriding tactical considerations, enemy air defense is always suppressed. The AATF FSO ensures that all flight routes and suspected enemy ADA sites are targeted with preplanned fires. Suppression of enemy air defense may be executed either as scheduled fires based upon a specific time schedule, or may be fired "on call," based upon the movement of the AATF through predetermined zones or across predetermined phase lines. The FSO is normally located with the AATFC and requires a dedicated fire direction net; he will control the lifting and/or shifting of SEAD fires as directed by the AATFC. Attack helicopter elements providing air assault security will suppress enemy ADA encountered en route. The attack battle team captain should select overmatch positions or fly escort along the flight route to provide immediate suppressive fires. For a complete discussion of these techniques, see FM 6-20.

i. The fire support plan may include any of the following categories of fire designed to complement the AATFC's operation:

(1) Planned fires on known or suspected enemy locations, avenues of approach, supply routes, and suspected weapons locations.

(2) On-call fires (prearranged fires that are requested).

(3) Preparations.

(4) Counterpreparations.

(5) Counterfires.

(6) Artillery delivered smoke (obscuration or screening).

(7) Illumination.

(8) Suppression fire.

(9) Scatterable mines (FASCAM can be delivered only when 155-mm howitzer systems are available to the AATF).

5-9. Coordination.

a. Effective fire support depends on decentralized execution and coordination. Based upon the AATFC's intent for using fire support, the FSO and FISTs execute the plan during the operation with minimum specific instruction. The FSO's coordination includes all actions required to make the plan work. He:

(1) Ensures the DS battalion FDC, the mortar platoon FDC, and any other supporting elements have the correct fire support plan and understand their portion of it.

(2) Verifies that the AATF mortars are in position to support, if available and required.

(3) Keeps fire support representatives at higher headquarters and the supporting field artillery TOC informed of the current tactical situation.

(4) Selects fire support means to attack targets during the operation.

(5) Keeps the AATF commander and S3 informed of the current status of fire support means available to the AATF.

(6) Recommends modifications of the fire support plan (during the operation) to react to battlefield changes, and ensures FISTs are aware of changes.

(7) Recommends, to the AATFC and/or S3, fire support coordinating measures to facilitate the attack of targets or to provide troop safety.

(8) Coordinates requests for additional fire support with higher level fire elements.

(9) Monitors execution of the fire support plan.

b. The FSO ensures that the plan developed remains supportable and must immediately inform the AATFC if there is not enough fire support allocated to make the plan work, or if changes are dictated in the plan. To do this, he is forward with the command group during the conduct of the operation. He normally flies with the AATFC when a C^2 helicopter is used.

c. The FSO keeps abreast of the tactical situation and coordinates all fire support impacting in his zone, including that requested by the AATF. He ensures that fires do not jeopardize troop safety, interfere with other fire support means, or disrupt adjacent unit operations. In this coordination, the FSO can utilize fire support coordinating measures.

d. During the conduct of the operation, shifts in priorities of fire, changes to the fire plan to support a changed scheme of maneuver, and immediate CAS are all handled forward by the FSO and ALO with the command group. The FSE at the TOC continues its planning responsibilities and provides backup support to the command group.

e. TheFSO,inconjunctionwiththeS3Air,coordinatesthefirecontrolactivitiesof the air assault task force (Figure 5-1).

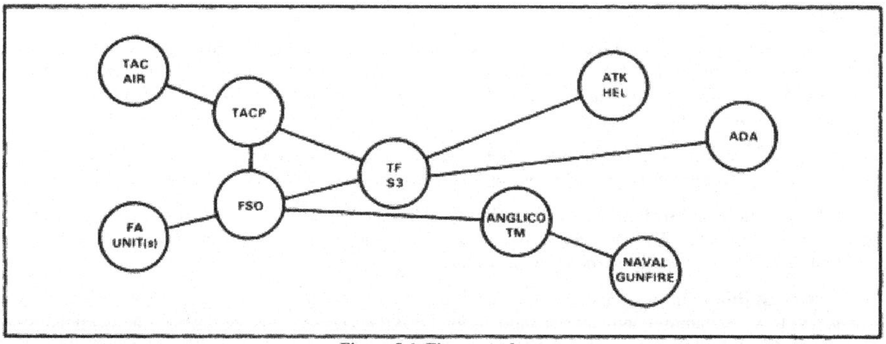

Figure 5-1. Fire control net.

f. Allaviatorsaretrainedtocallforandadjustindirectfires.Airreconnaissanceor attack helicopter unit aeroscouts may be particularly valuable in assisting the AATFC and the FSO in coordinating or adjusting indirect fires because they are normally in the best position to see the battlefield.

Section IV

ARTILLERY AIR ASSAULT OPERATIONS

5-10. General.

The AATF fights both offensive and defensive battles. Its organizational tactics, emphasizing aerial mobility and flexibility, require special planning considerations for employment. As discussed in Section III, the FSCOORD plans (with the AATFC) to support the ground tactical plan. In planning to support the air assault operations, the FSCOORD considers:

a. **Range for artillery and other fire support systems.** With the extended distances anticipated, the challenge for the FSCOORD is to position fire support systems so that they can range (place fire) and mass (concentrate fire) on targets within the AATF area of operations. When the AATF must operate out of artillery range, there is a greater dependence on CAS, attack helicopters, and mortars.

b. **Importance of the target.** Artillery is positioned to range those targets considered critical to the maneuver commander. For high value targets, the commander and the FSCOORD may consider moving artillery by helicopter to strike deep in the enemy's rear by firing across-FLOT raids or displacing laterally in sector.

c. **Airlift assets.** The mobility of the 105-mm direct support artillery battalion is one of its major characteristics. In taking advantage of its mobility to weight the operation, the commander must consider the cost of aircraft assets. To reposition the firepower of the DS battalion by air will normally require one aircraft (UH-60 or CH-47) per howitzer. The CH-47D can slingload two or more M1028 simultaneously. Additional aircraft must be committed to movement of vehicles and supplies necessary to support the mission.

d. **Risk in crossing lines.** A major consideration in planning air assault artillery operations is the risk in crossing enemy lines. The value of the target is weighed against the chances of survivability. Once the risk of crossing lines is considered, the FSCOORD must evaluate the survivability of the unit on the ground and during extraction.

e. **Target location.** For air assault operations, accurate LZ and target locations are essential. Accuracy of locations determines accuracy of fires and often targets will be engaged with unobserved fires.

f. **Pickup zone and/or landing zone.** Artillery displacements require PZs and LZs large enough to position equipment. When the unit arrives at the LZ, it must be secured and capable of supporting the unit that will most likely use the LZ as a firing point.

g. **Ammunition.** The amount of ammunition to be made available has a major impact on artillery support. When planning indirect fire support, the FSCOORD must consider the amount of ammunition required and the availability of transportation assets. Artillery ammunition supply operations will place a significant burden on aviation assets available to the AATF.

h. **Communications.** In the employment of field artillery, the ability to maintain communications is a requirement. The supporting unit must be within radio range of the supported unit to receive the call for fire (this is of particular concern when positioning the M198 with its maximum range of 30 kilometers). Unless unavoidable, the firing batteries must be within communications range of their parent battalion.

i. **Security.** The AATF artillery must rely on either terrain positioning or attachment of infantry to provide for security. The need for security forces is essential when FA units accompany the AATF across-FLOT.

5-11. Capabilities.

The artillery supporting the AATF should be organized with capabilities to match the needs peculiar to air assault operations.

a. The type of howitzer likely to participate in air assault operations is the towed 105-mm howitzer. Characteristics of the weapon, and 105-mm-equipped units, are listed below:

(1) **Responsive.** Capable of a high rate of fire.

(2) **Lightweight.** Capable of external slingload by the UH-60 and CH-47 helicopters.

(3) **Easily sustainable.** Towed artillery is less prone to downtime because it is not tied to a self-propelled carriage. It has reduced logistics requirements for a prescribed load list (PLL) and petroleum, oil, lubricants (POL) than self-propelled artillery.

(4) **Lack of crew protection.** The crews of towed howitzer batteries are especially vulnerable to direct and indirect fire. With no armor protection, the battery can expect heavy losses if engaged by the enemy.

(5) **Position security.** The mission of the field artillery is to provide indirect fires. FA is not designed with the capability to defend itself against a significant threat that may be encountered in across-FLOT air assault operations.

(6) **Limited range.** The 105-mm howitzer has a maximum range of 11,500 meters (15,100 with rocket assisted projectiles). The AATF can rapidly outdistance its supporting artillery.

(7) **Caliber.** The 105-mm howitzer is the smallest caliber howitzer in the Army inventory with a shell-burst radius of 35 meters.

(8) **Ammunition.** The 105-mm howitzer is limited to conventional munitions (high explosive [HE], illumination [ILLUM], improved conventional munitions [ICM], white phosphorus [WP], smoke [HC]) with limited chemical capability. The FASCAM is not available except to 155-mm howitzer equipped units.

b. TheM198howitzermayalsobeavailableforsupportofairassaultoperations.It is a 155-mm towed howitzer with a maximum range of 18,100 meters (30,000 meters with rocket assisted projectiles [RAP]). The M198 is movable by CH-47 C&D model aircraft. The 155-mm has a greater versatility in ammunition (HE, ILLUM, ICM, WP, HC, remote antiarmor mine system [RAAMS], area denial artillery munition [ADAM], dual-purpose improved conventional munitions [DPICM], chemical, nuclear [NUC]) with a bursting radius of 50 meters.

Section V

NAVAL GUNFIRE SUPPORT

5-12. General.

When operating near a coastline, naval gunfire support may be available to the AATF. Naval guns can provide high-volume, long-range, and accurate fires employing a variety of ammunition.

5-l3. Air and naval gunfire liaison company.

The ANGLICO provides ship-to-shore communications and fire control teams to adjust fire. In the absence of ANGLICO fire control teams, the AATF FISTS, aerial field artillery forward observers, or attack helicopter unit aeroscouts, may call for and adjust fires through the AATF ANGLICO team. See Appendix F, FM 6-20 for discussion for naval gunfire support.

Section VI

UNITED STATES AIR FORCE TACTICAL AIR SUPPORT

5-14. General.

The USAF support for the AATF normally includes tactical air reconnaissance, close air support, and tactical airlift. The AATF staff, in coordination with the air liaison officer, plans, integrates, and coordinates the Air Force support for air assault operations. Requests for air support are processed and discussed in Appendix E, FM 6-20.

5-15. Joint air attack team.

a. The joint attack air team (JAAT) is a combination of US Army attack helicopters and US Air Force close support aircraft (normally A-10 jet tactical aircraft) operating together to locate and attack high priority, lucrative targets. The JAAT normally operates in concert with field artillery or mortars, air defense artillery, and ground maneuver forces. Information flowing between the AATFC, the attack helicopter team leader, and the forward air controller (FAC) optimizes the effectiveness of attack helicopter teams and attack fighter flights in destroying the enemy force.

b. The JAAT can provide the AATFC with a highly mobile, extremely lethal tank-killing force capable of engaging enemy forces beyond the range of other antitank weapons. The JAAT can destroy or disrupt enemy formations and provide vital intelligence about enemy strengths and locations. By simultane ously employing attack helicopters and A-10s against the same target array at the same time, the AATFC increases the lethality and survivability of both systems.

c. When a JAAT is approved and is in direct support of an AATF, it is controlled by the AATFC. Otherwise, it is controlled by the attack helicopter commander or team leader who reports to brigade headquarters.

Section VII

AIR DEFENSE

5-16. General.

In the mid- to high-intensity environments, air assault operations normally require either local air parity or local air superiority. Since the number and type of air defense systems that can accompany the AATF is limited, and because helicopters are vulnerable to attacking aircraft, a great reliance must be placed on friendly air forces for air defense protection. Additionally, the AATF must optimize the employment of organic air defense weapons and maximize the use of passive defense measures.

5-17. Air defense standard tactical missions.

a. Air defense artillery unit missions are assigned using ADA standard tactical missions. These missions are much like those assigned the field artillery (to include support responsibilities for an ADA unit). They also establish support relationships to the supported unit or to another ADA unit.

b. The missions are general support, general support-reinforcing, reinforcing, and direct support. However, an AATF normally receives an ADA element in DS for close and continuous support. The ADA unit leader positions his weapons as necessary to properly support the AATF. The ADA may be attached, for movement, to infantry in order to facilitate control and security.

5-18. Air defense system.

Air defense protection for the AATF (within friendly lines) is provided by TACAIR and all elements of the ADA systems (Hawk, Patriot, Chaparral, Vulcan, Redeye, and Stinger). When the AATF penetrates enemy-held territory, air defense comes from ADA assets that can be displaced by helicopter. Due to weight restrictions, air defense forward of the FEBA is limited to organic Redeyes, Stingers, and Vulcans (towed). Although the towed Vulcan can be moved by UH-60, the prime mover must be displaced by CH-47. The Vulcan battalion has FM radios, and due to the extended distances between the Vulcan battery command post and the platoon elements (when it is attached to the AATF), the battery may require long-range AATF communication facilities.

5-19. Control and communications.

Air defense fire is controlled using the rules of engagement (determining type of aircraft and whether it is friend or foe) and weapons control status established by higher headquarters. Team leaders are responsible for deciding whether an aircraft is hostile or friendly. Weapons control status describes the relative degree with which the fires of air defense (AD) systems are managed. They are:

a. **Weapons free.** They may fire at aircraft not positively identified as friendly.

b. **Weapons tight.** They fire only at aircraft positively identified as hostile according to announced hostile criteria.

c. **Weapons hold.** They do not fire except in self defense or in response to formal orders.

5-20. Air defense priorities.

a. Priorities for air defense within the AATF are established by the AATFC. The senior air defense officer provides advice and makes recommendations based on his analysis of his area of operations to include the terrain, the high performance aircraft, attack helicopter avenues of approach, and all of the assets within his area of operations. The AATFC determines his priorities based upon:

(1) The AATF mission.

(2) How critical the asset or unit is to the accomplishment of the AATF mission.

(3) How vulnerable the target is to air attack.

(4) How quickly it can resume operations after it has been attacked.

(5) The enemy's ability to attack the asset.

(6) Coverage provided by other air defense systems.

b. Air defense priorities must be established for all periods to include before and after an operation, and would typically include protection for:

(1) Helicopter laager and assembly areas.

(2) Helicopter refueling and rearming points.

Section VIII

ENGINEER SUPPORT

5-21. General.

Combat engineers are an integral part of the combined arms team. Engineers possess the skills and equipment necessary to enhance friendly mobility and survivability, to counter the mobility of opposing forces, and to accomplish general engineer work. The engineers provide technical expertise and special equipment; the maneuver unit normally provides the required manpower.

a. **Categories of support.** Combat engineers provide four categories of support: mobility, countermobility, survivability, and general engineering.

(1) **Mobility.** Engineers reduce or eliminate the effects of obstacles to improve movement of maneuver forces and critical supplies. In support of air assault operations, engineers assist mobility by constricting or expanding helicopter LZs, FARPS, low altitude parachute extraction system (LAPES), and landing strips, and by maintaining, repairing, and rehabilitating existing forward aviation maintenance sites.

(2) **Countermobility.** Engineers construct obstacles to reinforce terrain to delay, disrupt, and kill the enemy. Countermobility increases time for target acquisition and maximizes the effectiveness of direct and indirect fire systems.

(3) **Survivability.** This involves protective position development: developing earth berms, dug-in positions, and overhead protection to reduce the effectiveness of enemy fire. In air assault operations, this could include protection of aircraft and fuel facilities.

(4) **General engineering.** These engineer missions do not directly contribute to the mobility, countermobility, and survivability of committed maneuver units. They are, however, essential for logistic support to include construction, improvement, and maintenance of rear area airfields.

b. **Command and support relationships.** The preferred engineer support relationship is DS; however, to conduct the air assault, engineers should be attached for movement only. During movement, engineers should be organized into squad-size elements and integrated into the air movement of infantry units. Once the movement is completed, the engineers should revert to DS and be task-organized no lower than platoon level.

5-22. Employment of engineer assets.

a. Engineer allocation to the brigade depends on METT-T, but will commonly be one company from the divisional combat engineer battalion. When requirements exceed the capabilities of one company, additional resources from either the division engineer battalion or support corps engineer units may be made available.

b. At the task force level, the number of engineer personnel and their relationship (to the TF) is dependent on METT-T. Even if no support relationship is established, engineers may have assigned missions in the AATF area and coordination must be maintained. This coordination is best effected by detailing an officer and a noncommissioned officer (NCO) from the supporting engineer company to the AATF for the duration of the operation.

c. How to use available assets is an important planning function.

(1) **Plan and prioritize.** The use of a scarce resource, such as combat engineers, must be carefully planned. The AATF engineer is part of the planning process from the beginning. The AATFC, S3, and the engineer work together to plan the use of the engineer assets and establish priorities. The engineer then advises the commander on how best to utilize assets based on time, personnel, equipment, and munitions available. A clear list of priority tasks is determined based on the AATFC's guidance and the engineer's recommendation.

(2) **Integrate.** The AATF commander ensures that the engineer effort is integrated into the scheme of maneuver and fire support plan. Fires, both direct and indirect, are planned to cover all obstacles. An obstacle placed where it cannot be covered by fire is a wasted effort. Therefore, the S3 ensures that he includes the FSO and engineer together in his planning effort. In addition, the AATFC and S3 direct the integration of AATF personnel into the accomplishment of all engineer work. Engineer assets must be placed well forward in the scheme of maneuver to assist the mobility of maneuver forces in the critical, early stages of attack.

(3) **Control.** The AATFC supervises the accomplishment of the engineer's mission as he prioritized it. Changes in the situation may require changes in the priority of engineer work and the AATFC and/or S3 communicate such changes to the engineer.

(4) **Support.** Mobility, countermobility, and survivability tasks are the responsibility of the AATF, not the engineer. While the engineer unit will provide much of the manpower allocated to these functions, use of other elements is normal. An example is the use of infantry to construct obstacles under the supervision of an engineer. To allow all engineer assets to be used for engineer tasks, AATF combat elements usually provide security for the engineers. The CSS for the engineer unit is provided by the parent engineer unit, except when engineers are attached. Regardless of the command and support relationship, the AATF provides Class IV and Class V stores to support its engineer operations.

(5) **Execute.** The engineers accomplish their mission in support of the AATFC's scheme of maneuver.

d. Special considerations involve heavy equipment found in engineer battalions, other than air assault or airborne units, which is generally too heavy to be air assaulted. Plans must be made to link up with the equipment later in the operation. If air movable heavy equipment is available to the AATF, their capability is greatly increased; however, resupply of diesel fuel becomes an important planning consideration. Another consideration is the special equipment necessary to slingload engineer equipment (chain leg sets and A-22 bags). This equipment is not normally available from the aviation units but will be provided by the engineer units possessing air movable heavy equipment. When air movable heavy equipment is not available, engineers must be prepared to construct obstacles, barriers, and assist with survivability tasks, by employing hand tools, explosives, and field expedient methods.

CHAPTER 6

Combat Service Support

Section I

PLANNING

6-1. General.

Combat service support for air assault operations must be planned, organized, and executed to support a rapid tempo in highly mobile and widely dispersed operations. The traditional doctrinal distances and responsiblities do not always apply to air assault operations. The air assault logistical planner must recognize this from the outset and be prepared to adapt and innovate with the resources at hand. Just as the AATF is tailored for combat operations by air, the logistical system must be tailored to support by air and is therefore dependent upon considerable outside support.

6-2. Air assault task force S4.

a. It is imperative that the AATF S4 be involved in the planning of air assault operations from the initial stages onward. This ensures that all facets and constraints of logistical support are considered, and provides the lead time necessary to organize and position those units and resources that are required to support the mission.

b. The S4 and the S3 must coordinate closely and continuously throughout any air assault operation. Both must share the same resource for moving combat power and sustaining assets - the helicopter.

c. To organize CSS for air assault operations, the logistical planner must know:

(1) The task force mission.

(2) The concept and duration of the operation.

(3) The task organization to include densities of-

(a) Personnel.

(b) Weapon systems by type.

(c) Equipment by type.

(d) Aircraft by type.

d. He must also consider the impact of the following:

(1) Enemy situation.

(2) Weather.

(3) Terrain.

(4) Reliance on air lines of communications.

(5) Great distances between supporting and supported units.

(6) Large ammunition and aviation fuel consumption rates.

e. The prudent air assault logistical planner will ensure that CSS is provided, not only for his organic and attached elements, but also for DS or OPCON units. Although the AATF does not have the inherent responsibility for CSS to OPCON, DS, or GS units, it does have the responsibility to ensure that CSS is coordinated for the supporting unit(s). The coordination expressly designates who provides CSS throughout the operation. When a large attachment joins the AATF, the attachment should bring appropriate amounts of its own CSS assets from its parent unit. These assets are controlled by the AATF administrative logistics center to provide coordinated CSS to the attached unit.

Section II

COMBAT SERVICE SUPPORT ORGANIZATION

6-3. General.

a. The AATF is supported by both organic and external elements organized to push supplies, materiel, fuel, and ammunition forward by air.

b. A brigade-size AATF must rely on the forward area support team (FAST) or the forward support battalion (FSB) to support the operation. When the AATF is organized around an infantry battalion nucleus, a task-organized forward service support element may be dedicated to supporting the air assault operation.

c. A battalion AATF FSSE would typically consist of.-

(1) Medical support - a light shocks section.

(2) Maintenance contact teams for communications, automotive, armament, and recovery.

(3) Class III (ground) and Class V support sections.

d. The exact organization and disposition of CSS elements is a function of the AATF's mission and anticipated follow-on operations. Normally, two options for organizing and positioning CSS elements prevail:

(1) If the AATF anticipates being extracted from the objective area following mission completion, unit trains and supporting CSS elements are not normally displaced forward but remain in the brigade support area (BSA) or other rear area.

(2) If the AATF is to remain in the air assault objective area to link up with other forces, or to conduct extensive follow-on operations, CSS elements would initially be provided by moving FSSEs and combat trains forward when the enemy situation permits.

6-4. Air assault task force trains.

a. The trains for all AATF elements must be organized, located, and controlled so as to facilitate the consolidation, packaging, and air movement of support packages configured to unit size (normally company or platoon).

b. The organization of trains varies with the mission assigned the AATF and the CS and CSS available. Trains may be centralized in one location (unit trains), or they may be echeloned in two or more locations (echeloned trains). It is normally appropriate to centralize all AATF logistical assets at one location as unit trains under the control of the AATF S4. This provides ease of coordination, control, and security of logistical assets, and allows for the most efficient use of logistical support helicopters.

c. Echeloned trains are normally only used when operations extend over vast distances such as might be expected in the delay or during an economy-of-force or security mission. The AATFC would normally elect to echelon his trains when he feels that CSS must be collocated with maneuver units to provide immediate, dedicated support.

d. The AATF commander normally moves only essential support elements to the objective area.

6-5. Command, control, and communications (C3).

a. The AATF SI and S4, under the direction of the AATF executive officer (XO), operate the administrative and logistics center and have overall responsibility for CSS command and control.

b. Timely and effective CSS depends on a good communications system. At AATF level, CSS communication may be by radio, courier, or radio teletypewriter (RATT). The AATF administrative and logistics center radio net is used for most administrative and logistic traffic. For lengthy administrative and logistical reports, messenger or RATT should be used. "As of" and "due" times for reports at all levels should take this into consideration, allowing more time for long reports to be delivered by messenger.

c. The administrative and logistics center is the net control station (NCS) for the administrative and logistics net. The S4, SI, headquarters company commander, maintenance officer, support platoon leader, medical platoon leader, company first sergeants, and others as required, operate in the administrative and logistics net.

d. When FM radio communication over the AATF administrative and logistics net is not possible, due to the distance between stations, hard copy messages are sent with resupply or evacuation aircraft.

6-6. Helicopter external load operations.

a. Transporting supplies and equipment by helicopter external (sling) load has the advantage of rapidly moving heavy, outsized, or urgently needed items directly to the using unit. The logistical planner can enhance the sustainment of the AATF by planning well in advance for slingload operations and by understanding the limitations imposed by external load operations.

b. External load limitations to be considered:

(1) If a cargo is too light or bulky, it will not "fly" properly when suspended under the aircraft at cruise airspeeds.

(2) The external load must not exceed a helicopter's lift (under given atmospheric conditions) or hook capabilities (8,000 pounds for the UH-60).

(3) Airspeeds must be slower when helicopters carry external loads.

(4) Dust, sand, or snow, which would be blown during hover for pickup or delivery of cargo, may preclude safe external load operations.

(5) The higher altitudes, which must be flown with slingloads, may subject the aircraft to more ground fire.

(6) Extended hovering to pick up or deliver a slingload during darkness is inherently more dangerous than similar daylight operations.

(7) The availability of suitable slings, cargo nets, cargo bags, and other air delivery items may preclude or limit external load operations.

c. There are normally three different elements involved in a slingload mission: the supported unit, the aviation unit, and the receiving unit. The responsibilities and functions of each are:

(1) The supported unit (normally the AATF S4) is responsible for:

(a) Selecting, preparing, and controlling the PZ.

(b) Requisitioning all the equipment needed for slingload operations, including slings, A-22 bags, cargo nets, and containers.

(c) Storing, inspecting, and maintaining all slingload equipment.

(d) Providing a sufficient number of trained ground crews for rigging and inspecting all the loads, guiding the helicopters, hooking up the loads, and clearing the aircraft for departure.

(e) Securing and protecting sensitive items of supply and equipment.

(f) Providing load derigging and disposition instructions to the receiving unit.

(g) Providing disposition instructions to the receiving and aviation units for the slings, A-22 bags, cargo nets, and containers.

(2) The aviation unit is responsible for-.

(a) Effecting and/or establishing coordination with the supported and receiving units.

(b) Advising the supported unit on the limitations of the size and weight of the loads that may be rigged.

(c) Advising the supported and receiving units on the suitability of the selected PZs and/or LZs.

(d) Providing assistance for the recovery and return to the PZ of the slings, A-22 bags, cargo nets, and containers as required by the supported unit.

(e) Establishing safety procedures that will ensure uniformity and understanding of duties and responsibilities between the ground crew and flight crew.

(3) The receiving unit is responsible for:

(a) Selecting, preparing, and controlling the LZ.

(b) Having trained ground crews available to guide the aircraft in and derig the load.

(c) Coordinating with the supported (sending) unit for the control and return of the slings, A-22 bags, or any other items that belong to the supported unit, and returning them as soon as possible.

(d) Preparing, coordinating, and inspecting backloads, such as slings, A-22 bags, and so forth, and having them ready for hookup or loading.

d. See FM 55-450-1 for additional information on these procedures.

Section III

COMBAT SERVICE SUPPORT EXECUTION

6-7. General.

The AATF is normally configured to conduct the initial assault with one to three days of accompanying supplies to ensure some degree of self sustainment. When the enemy situation permits, resupply is accomplished by air on a routine basis to keep supplies at the one- to three-day level.

6-8. Supply.

a. The most efficient method for conducting the resupply of forward AATF units is the logistics package (LOGPAC) method. A LOGPAC is a resupply element based on a day's logistics requirements for a company. It is organized in the unit trains by the company supply sergeant and the support platoon leader and prepared for air movement. The AATF SOP establishes the standard LOGPAC.

b. Supplies going forward from the trains move by methods that reduce loading and unloading times. Palletized or external slingloads reduce the ground-time vulnerability of aircraft because they can be unloaded quickly.

c. When preparing the loads, the S4 provides essential equipment and personnel (for example, hookup teams, ground guides, signalmen, slings, pallets, nylon webbing, chainlink slings, and clevises).

6-9. Maintenance.

Maintenance involves inspecting, testing, servicing, repairing, requisitioning, rebuilding, recovering, and evacuating. The assault echelon is not normally accompanied by maintenance personnel. During air assault operations, repair above the operator level is accomplished in one of two ways:

a. Contact teams from organizational or support maintenance may be flown forward to effect immediate repair of critical equipment.

b. Deadlined and/or damaged equipment is evacuated by air.

6-10. Field and personnel support services.

These services for the soldier are an important part of the overall support effort and continue during air assault operations; however, these services are rarely part of an air assault operation. Rather, they are accomplished in a rear area outside the air assault area of operations.

6-11. Medical support.

a. **Support.** This is provided by the medical platoon and the medical section of the FSSE, when available. The support is planned by the AATF medical platoon leader and is addressed in the administrative and logistics annex to the OPORD to include:

(1) Location of far forward treatment sites.

(2) Ground and air evacuation plans and/or routes.

(3) Location of support hospitals.

(4) Communications instructions.

To adequately support the mission, the platoon leader should be included in all operational and/or tactical briefings.

b. **Medical evacuation.** The primary means for AATF medical evacuation is by medevac helicopter. Inflight medical care is essential for those patients whose condition is serious enough to require air evacuation. Medical evacuation crews will deliver patients to proper treatment facilities.

c. **Control.** There are two options for controlling medevac requests:

(1) Allow subordinate units to request medevac direct from the medical unit.

(2) Receiving and consolidating requests, establishing casualty priorities, and dispatching medevac aircraft.

d. **Coordination.** When possible, the AATF SI coordinates directly with the medevac unit commander or section leader. He provides the unit a complete copy of the AATF's CEOI, PZs, LZs, and flight route overlay. This makes it possible for the medevac helicopter pilot to establish radio contact on the internal radio net of the supported unit. This helps relieve congestion on the command radio net. When medevac communications take place at the AATF level, they are usually done on the administrative and logistics net. Since the medevac unit has the PZs, LZs, and flight routes, the AATF commander can direct medevac aircraft through his sector via a specific flight route. This lessens interference with ground operations, indirect fires, and TACAIR. A technique used in the employment of medevac helicopters is to have them trail the AATF while it is en route. This ensures that the helicopters are immediately available to take on wounded and ensures pilot familiarity with the route to the objective area. If evacuation is required later, faster response is possible.

e. **Nuclear, biological, and chemical.** In the event that contaminated casualties have to be evacuated by helicopter, the aircrew should be warned before actual pickup. When flying with these casualties, all personnel on the aircraft must be in appropriate mission-oriented protection posture (MOPP) and land downwind from the medical aid station after notifying the station of casualty status and the contaminating agent.

6-12. Transportation.

In addition to their assault and assault support roles, utility and medium (CH47) helicopters play an important role in providing CSS for the AATF. Helicopters are relied upon for movement of supplies, materiel, fuel, ammunition, maintenance contact, and for evacuation of damaged equipment.

Section IV

AVIATION SUPPORT CONSIDERATIONS

6-13. General.

Aviation units consume large amounts of fuel, ammunition, Class IX, and maintenance support during intensive air assault operations. Although aviation units are normally responsible for meeting their own unique logistical support requirements, the air assault logistical planner must be aware of the requirements, plan for them, and be prepared to assist as necessary.

6-14. Forward area rearming and refueling points.

Forward area rearming and refueling points are established by aviation units to provide for the rapid rearming and refueling necessary to sustain a fast pace. Forward area rearming and refueling points are:

a. Established in the vicinity of the ground unit exercising operational control (behind the FEBA, and out of range of enemy artillery).

b. Positioned to reduce turnaround time, thus optimizing helicopter availability, and repositioned frequently to avoid detection and destruction.

c. Fully mobile, using ground vehicles and helicopters.

d. Capable of operation within 30 minutes of installation and capable of redeployment within 30 minutes.

e. Capable of performing refueling and rearming operations rapidly and efficiently.

6-15. Aircraft maintenance and recovery.

a. **Maintenance.** Aircraft have substantial maintenance requirements. However, maintenance is kept to a minimum in the operational area. A method used to accomplish this, and still have responsive maintenance, is to have aircraft standing by to move maintenance contact teams where required. In addition, a maintenance aircraft and personnel may accompany the flight. If an aircraft has maintenance problems during movement, maintenance personnel may be able to repair it and save valuable aircraft operation time.

b. **Recovery.** If an aircraft is forced to land on enemy terrain due to mechanical problems or combat damage, every effort is made to protect the aircraft and crew until they can be evacuated. However, mission execution has priority over rescue and recovery operations. The AATF commander is notified immediately of any downed aircraft. He takes action to secure and recover the crew and aircraft with his resources or requests recovery by higher echelon. When an aircraft is downed, the senior occupant assumes command and establishes a defense of the area or organizes evasive action. If an aircraft is abandoned, steps are taken to destroy it to preclude its capture or the capture of sensitive equipment or documents. The level of authority required to destroy the aircraft is established in higher echelon SOPs (it may be covered in the OPORD). However, if capture is imminent, the aircraft, equipment, or documents should be destroyed. Recovery of a downed aircraft is accomplished by the aviation unit. The AATF commander may have to provide security for the recovery team.

APPENDIX A

Conduct of an Air Assault Operation

Section I

CONCEPT

A-1. General.

The following sequence demonstrates the air assault task force movement from the assembly area to the pickup zone and on to the landing zone. Some steps may be deleted when not appropriate, or they may be accomplished concurrently with other steps.

A-2. Movement.

The AATF operation normally begins at the assembly area.

Section II

MOVEMENT FROM ASSEMBLY AREA TO LANDING ZONE

A-3. General.

a. At the prescribed time, units move from the assembly area to the holding area, via a route designated by the AATFC. A holding area must be:

(1) Covered and concealed.

(2) Sufficient size for the assault force,

(3) Close to primary and alternate PZs.

b. Each unit commander notifies the PZ control party upon his unit's arrival in the holding area. In this area, unit leaders separate the unit into loads according to the loading plan. Heavy loads and slingloads should not be programmed in initial serials. Offloading heavy internal loads is time-consuming and slows troop buildup (Figure A-1).

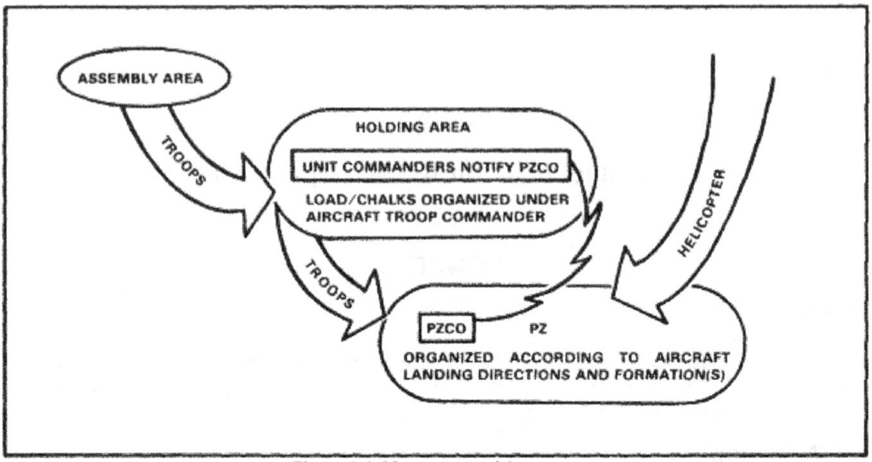

Figure A-1. Movement to pickup zone.

c. Each load includes a designated aircraft troop commander responsible for briefing his troops and inspecting the load. He ensures that the load is organized and ready to be loaded as planned. The PZ control party briefing includes the loading point for primary and alternate PZs and the routes to those points. The aircraft troop commander briefs his personnel. As a minimum the briefing includes:

(1) Loading procedures.

(2) Bump plan (for individual and/or load bumps).

(3) Use of safety belts.

(4) Preflight safety inspection of soldiers.

(5) In-flight procedures.

(6) Downed aircraft procedures

(7) Offloading procedures.

(8) Movement from the LZ.

A -4. Organization of the pickup zone.

a. The PZCO lays out the PZ as directed in the plan. For example, if the plan calls for landing to the west in a staggered trail formation with the lead aircraft landing on a spot marked by a smoke grenade, or panel markers, the PZ is laid out that way. Pathfinders should assist the PZCO when available (Figure A-2)

Figure A-2. Organization of the pickup zone.

b. It is imperative that aviation elements arrive at the PZ in the formation directed in the plan. This minimizes confusion during loading. The PZCO, or pathfinder element, assists in loading by ensuring aircraft and personnel are in the proper location and formation at the correct time. If an aircraft (scheduled for the lift) is unable to complete its mission due to mechanical failure, the PZCO informs the unit commander, who implements the aircraft "bump" plan.

A -5. Infantry movement to pickup zone.

a. The PZCO coordinates the arrival of aircraft and troops so that the troops arrive at their respective loading points just before the aircraft land. This prevents congestion, facilitates security, and reduces vulnerability to enemy actions during PZ operations.

b. On the PZCO's signal, aircraft troop loads move by designated routes from their holding area(s) to their loading points on the PZ. The PZCO may use schedules, messengers, arm-and-hand signals, light signals, or (as a last resort) radio to order helicopter loads to move to the PZ.

c. If the primary PZ cannot be used ,the PZCO advises the unit commander to move to the alternate PZ.

A -6. Helicopter movement to pickup zone.

a. Aircraft begin movement to arrive on the PZ at the time listed on the air movement table. The PZCO contacts the aviation element if there is a PZ change. If there has been a change in allowable cargo load, number of aircraft, or formation, the AMC contacts PZ control.

b. During air movement to the PZ, enemy antiaircraft or other fire may be encountered. Therefore, air cavalry scout teams may be used to locate and suppress enemy positions prior to the arrival of the lift aircraft. Air cavalry scout teams may also be employed on the flanks and to the rear of the lift aircraft. Attack helicopter and/or air cavalry scout teams will not normally land on the PZ. When the lift helicopters are to be on the ground for extended periods, the attack helicopter, air cavalry teams may occupy holding areas nearby or return to rearm-refuel sites. The command-and-control helicopter is positioned where the command group can see and control critical events.

c. Strict radio discipline is maintained throughout the operation; radio silence should not be broken unless absolutely necessary. Radio calls between aircraft are permitted only as a last resort when other signals are not appropriate.

d. The helicopters should use terrain flying techniques en route to the PZ.

A -7. Lift-off from the pickup zone.

a. When the aircraft are loaded and ready for lift-off, the PZCO signals the flight leader using arm-and-hand or light signals. The flight leader may signal other aircraft by turning on (or off) his navigation lights. For example, upon landing, the lights are turned on, and when they are turned off, the flight lifts off. Members of the PZ control party may also relay the alert to lift off to aircraft in the rear of the formation, or the flight leader simply lifts off and the others follow (Figure A-3).

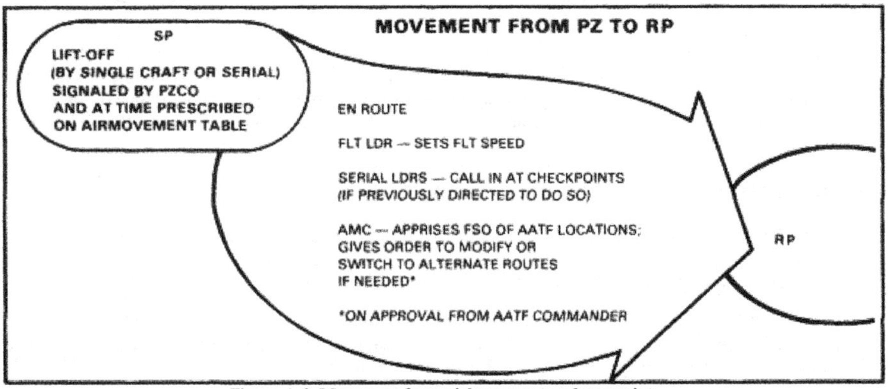

Figure A-3. Movement from pickup zone to release point.

b. Lift-off should be at the time prescribed in the air movement table. However, the aircraft will not loiter in the PZ. If they are early, they lift off and alter speed to cross the SP or first ACP on time. This should place the first aircraft of the first lift in the LZ at H-hour.

c. Lift-off may be by single aircraft or by serial. Under some conditions (dusty PZ, restricted PZ, or high density altitude and no wind), it is best to break serials into smaller increments. When possible, simultaneous lift-off is best for the following reasons:

(1) It is easier for the attack helicopter unit commander to plan his scheme of maneuver and provide security en route for aircraft.

(2) AATF control is more positive.

(3) It reduces the enemy's time to fire at the aircraft.

d. Theflightleaderadjuststheflight'sspeedandrateofclimbsoallelementsform into the en route flight formation at the required altitude.

A -8. En route to the landing zone.

a. The AMC predetermines the en route flight speed and the flight leader paces the flight to ensure the flight crosses the SP on time.

b. Radio silence is paramount; however, if directed in the order, serial leaders report to the AMC on passing each communication checkpoint. En route radio calls are made only if the flight is late or if it is required to deviate from the plan.

c. Troop unit commanders, leaders, and aircraft troop commanders must remain oriented throughout the flight. They do this by following and verifying the flight route using terrain observation, maps, aircraft compass, and aircraft speed.

d. When a threat is encountered along the flight route, such as heavy enemy fire, the AATFC gives the order for the AMC to modify or switch to an alternate flight route. The AMC's radio traffic is brief when shifting aircraft to an alternate flight route. If the LZ is to be changed, the AATF commander makes the decision and informs the AMC. If the AATFC cannot be contacted, the senior ground commander in the flight will make the decision. It is recommended that the AATFC or an S3 representative fly with the AMC to facilitate command and control.

A -9. Security.

a. Air cavalry and attack helicopter units provide security for downed aircraft, route reconnaissance, and other assistance en route (as desired by the AMC).

b. United States Air Force aircraft (when assigned this role) may work with attack helicopter units to provide security to the flanks, front, and rear of the helicopter formations). When performing this role in a medium-to-high Threat environment, specially equipped aircraft suppress or destroy surface-to-air missile sites and radar-directed guns. Other USAF aircraft may be used to selectively jam enemy radar and communication signals using jamming transmitters or other methods such as "chaff '(dropping shredded aluminum foil strips to foul radar). Ground attack aircraft, A-10, A-7, A-37, or AC-130 may be with, or in advance of, the flight formation, or may be on alert nearby or in planned orbits and support patterns to respond rapidly.

c. Indirect fire weapons provide suppressive fires along the flight routes as planned or as necessary.

d. If a lift aircraft emergency occurs (forced landing in an unsecure area), the aircraft commander (if time permits) switches his radio to the "guard channel" and transmits a "Mayday" in the clear. He announces his identification, heading, position, nature of emergency, and intentions. (The aircraft crew alerts passengers of the emergency and secures loose equipment.) The SOP for downed aircraft is then put into effect. The AATFC makes the decision on whether the ground element aboard the aircraft moves to a linkup point and continues the mission or remains with the aircraft.

A-10. Landing operations.

a. Afterpassingthereleasepoint,serialsproceedtoassignedLZs.TheRPcrossing is used to time the lifting and shifting of artillery and close air support strikes, if preparatory fires are used. The RP is also the point at which the aircraft shift to LZ formation, if required (Figure A-4).

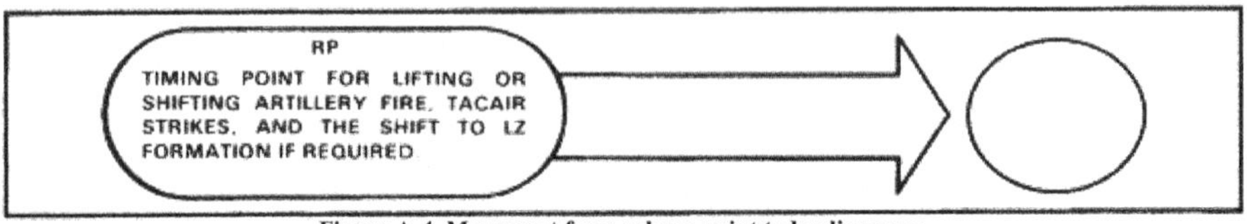

Figure A-4. Movement from release point to landing zone.

b. Napalm and other incendiary ordnance are not normally used on the LZ and its immediate vicinity (just prior to landing), because foliage fire and smoke could endanger aircraft or hamper the mission. However, helicopters equipped with smoke generators can be utilized to provide a smoke screen. Wind direction, speed, and enemy air defense must be considered along with friendly indirect fire support.

c. Attack helicopter units and/or teams are employed in various roles during an LZ operation. They may:

(1) Precede the lift element into the LZ (by a few minutes) for reconnaissance and/or to provide suppressive fires to prevent a time gap in LZ fires (provided by other support elements).

(2) Recommend last-minute changes regarding aircraft landing instructions.

(3) Provide area cover and neutralize known enemy positions, or provide security for lift aircraft while in the LZ area.

(4) Observe ground approaches to the LZ for possible enemy attacks.

NOTE: After the initial pass, attack helicopters may enter an overmatch flight pattern around the LZ.

A-11. Command and control helicopter.

At the RP, the command-and-control helicopter moves into position (employing terrain flying) to observe and communicate with assault elements. To avoid enemy weapons, the pilot uses popup techniques to observe activity. The AATFC will determine where he can best influence the action, by remaining on the aircraft or joining the ground forces.

A-12. Support fires.

a. Preparatory fires should be planned for all primary and alternate LZs. The decision to fire a specific LZ preparation is made by the AATFC. The FSO should travel with the AATFC to expedite fires and changes to preplanned fires. Fires will be planned along all routes leading to the LZ. Planned fire should be intense. Fires shift or lift shortly before the first assault element lands. A preparatory fire sequence might look as follows:

TIME (minutes)	ACTION
H-20	Recon by air cavalry completed.[1]
H-20 to H-5	Tactical air strike (USAF).[2]
H-5 to H-2	FA preparation.
H-2 to H-Hour	FA shifts fires; attack helicopter suppression.
H-Hour	First lift lands.
H-Hour to H+30	Attack helicopters .[3]
H+30 to H+120	Field artillery.[4]

[1]Conduct route reconnaissance (recon) from PZ to LZ. At H-20 move to alternate LZS and continue reconnaissance.

[2]FA and tactical air may engage simultaneously if sufficient ammunition is available.

[3]On station for targets of opportunity in the vicinity of the LZ

[4]On-call suppressive fires and counterbattery fires.

b. In the development and sequencing of fire plans, the following are considered:

(1) **Deception.** False preparations may be fired in areas other than the objective area.

(2) **Loss of surprise.** A preparation of long duration may reduce the possibility of surprise.

(3) **Availability of fire support.** The FSO considers the assets that can fire preparations and coordinates the arrangements with the FA battalion S3. Preparations by USAF tactical aircraft are requested through the FAC.

(4) **Significant targets.** A known or suspected enemy force located in the vicinity of the LZ, regardless of size, warrants LZ preparation.

(5) **Shifting fires.** Artillery fire continues throughout the assault phase, shifting from the LZ to known or suspected targets.

(6) **Obstacles to landing and maneuver.** Various types of ordnance used in a preparation can cause craters, tree blow-down, fires, smoke, and poor visibility on and near the LZ.

(7) **Positive control measures.** Control measures must be established for lifting or shifting fires; for example, restrictive fire line or restrictive fire area (RFA).

(8) **Ammunition.** Basic load and resupply limitations.

c. Because CAS station time is limited by fuel and enemy air defenses, the sequencing of support fire is carefully controlled by the FSO to obtain maximum, continuous support. To ensure that all fire support assets are utilized at the correct time, the FSO must be constantly informed as to the status of the flight. This allows him to orchestrate fires to coincide with the actual arrival of assaulting elements at the LZ (Figure A-5).

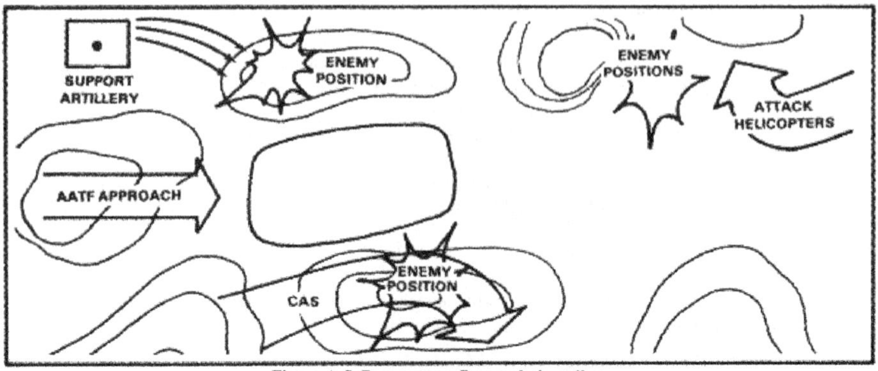

Figure A-5. Preparatory fires and air strikes.

d. Another method of continuing assault fire support is to shift indirect fires to one flank, conduct a simultaneous airstrike on another flank, and use the attack helicopter teams to orient on the approach and departure routes. This technique requires precise timing and assault formation navigation to avoid flight paths of other aircraft and gun target lines of indirect fire weapons.

A -13. Landing techniques.

a. The AATF lands as planned unless last-minute changes in the tactical situation force the commander to abort or alter the landing. The aviation crew must make every effort to keep the troops in their aircraft informed of the situation, especially of any changes to the original plan.

b. A simultaneous landing is desired so as to place the maximum number of troops on the ground, in a given area, in the shortest possible time. Individual aircraft touchdown points are planned to disembark troops as close as possible to their initial positions.

c. The operation is accomplished with a minimum number of lifts, each with the maximum number of aircraft the LZ will accommodate. This reduces the expo sure time of the aircraft, maintains unit integrity, provides maximum combat power, and gives the enemy less time to react. When separate element landings are dictated because of LZ size, time intervals between elements are kept as short as possible. Ideal timing has an aircraft element landing immediately after the preceding element lifts off.

d. Troops are most vulnerable during landing; they disembark rapidly and deploy to carry out assigned missions.

e. An air evacuation location is designated, normally at the approach end of the LZ. This permits continuation of the lift and prompt evacuation of the wounded.

f. At the LZ, leaders at each command level account for all personnel and equipment and submit appropriate reports to higher headquarters. Key personnel killed, wounded, or missing are replaced according to unit SOP; key weapons missing or out of action may require the force to reorganize.

After the unit completes its consolidation of the LZ, it is reorganized as necessary. Ground combat operations are no different from those conducted by other infantry units (FM 7-20).

A-14. Lift helicopter(s) return trip.

a. When the LZ operation *is* finished, aviation elements return by preselected routes to complete subsequent lifts or to conduct other operations or, if prescribed in the order, move to a laager.

b. If subsequent lifts are required in the same operation, the procedures described above are repeated.

Section III

AIR ASSAULT SCENARIO

A-15. General.

This is an example of air assault in a counterinsurgency situation, with map overlay schematics, an operation order, and an air movement table, describing a unit in action.

A-16. Basic situation.

a. The 21st Light Infantry Division, as a part of a joint task force, has been deployed to Palomas for an unspecified period of time. The deployment of the division was in response to a request for increased US assistance due to the deteriorating military and political situation within Palomas. The division will augment the Palomas Army and provide the needed additional forces for security operations until the Palomas Army is expanded. After completion of the deployment phase, the brigades set up support bases in their assigned area of operations. The 2d Brigade began conducting offensive operations in the state of Sedona.

b. On 10 June, the scout platoon from the 1st battalion located what is believed to be a guerrilla base camp. The enemy force concentration indicated they were preparing for an offensive operation within several days. The scout platoon stated the camp was occupied by an estimated reinforced platoon of 50 to 60 men armed with AK assault rifles, RPK machine guns, RPG rocket grenade launchers, and SA7 portable air defense missiles. The fleeting target would have to be attacked quickly or an opportunity would be lost. The 2d Brigade commander ordered the Ist Battalion to destroy the enemy force and the guerrilla base camp.

A-17. Special situation.

The 1st Battalion, 20th Infantry, is in an assembly area on the north side of the brigade operational support base. At 101100 June, the battalion received the brigade order to conduct an air assault operation to destroy the enemy force and the guerrilla base camp at 110530 June. The battalion immediately began preparing for the operation. The battalion commander issued a warning order to subordinate commanders and planning guidance to his staff which resulted in the formulation of an operation order. A copy of the written order follows:

SAMPLE OPERATION ORDER

____

(CLASSIFICATION)

AIR ASSAULT

Copy No ______of______ Copies
TF 120 INF
PJ
101100S JUN
JT21

OPORD 6-85 (EAGLE)

Reference: Map, series B142, Palomas Special Edition,

1:50,000

Time Zone Used Throughout the Order: Sierra

Task Organization:

Co A FIST/1-221 FA	Co B FIST/1-221 FA	Co C FIST/1-221 FA Mort Plt (for movement only)

TF Con

Scout Plt
A/21 Atk Hel Bn (OPCON)
TM/B/1-11 ACS (OPCON)
210 CAC (DS)
211 CAC (DS)
FSE/1-221 FA
TACP
AVN LO

1. SITUATION.

a. Enemy Forces.

(1) Elements of a reinforced platoon occupy positions vic (_________).

(2) The enemy is capable of employing dismounted weapons systems; it is unlikely that the enemy will have substantial artillery support.

(3) The enemy will initially defend but most likely avoid decisive engagement and attempt to escape along trails to the southeast and southwest from the guerrilla base camp.

b. Friendly Forces.

(1) 2d Bde conducts offensive operations from present location commencing _________ with the intent of destroying enemy forces in assigned area of operations.

(2) TF 2-20 conducts offensive operations from present location commencing _________ to establish company-size blocking positions _________ , and _________ .Oriented toward the southeast.

(3) TF 3-20 conducts offensive operations from present location commencing _________ to establish company-size blocking positions _________ , and _________ . Oriented toward the southwest.

(4) 1-221 FA DS to 2d Bde.

c. Attachments and Detachments: None.

2. MISSION. TF 1-20 air assaults at 110530 Jun to attack and destroy the enemy forces and the guerrilla base camp, located vicinity coordinates _________ .

3. EXECUTION.

a. Concept of Operation. Annex B (Operation Overlay)

(1) Maneuver. TF 1-20 conducts multiple company-size air assaults from PZSX (_________), Y(_________), and Z (_________) and attacks to seize Objectives RED, WHITE, and BLUE. Co A, Co B, and Co C air assault simultaneously commencing 110530 Jun, with Co A to LZ A (_________) to establish platoon-size blocking positions within OBJ RED (_________) to destroy any escaping enemy from the guerrilla base camp; Co B to LZ B (_________) makes the main attack to destroy the enemy forces and the guerrilla base camp (OBJ WHITE _________). Co C to LZ C (_________) to establish platoon-size blocking positions within OBJ BLUE (_________), to destroy any escaping enemy from the guerrilla base camp. My intent is for Co B to attack the guerrilla base camp to destroy it, killing or capturing as many guerrillas as possible. Co A and Co C will simultaneously be positioned along the routes of escape to set up platoon-size blocking positions to kill or capture escaping guerrillas. Annex C, Air Movement.

(2) Fires. Annex D, Fires Support Overlay.

(a) Priority of FA fires initially to Co B, then to Co A.

(b) A 3-minute prep will be fired on LZ B from H-5 to H-2.

(c) Priority of mortar fires, upon completion of air assault, initially to Co B, then to Co C.

b. Co A:

(1) Conduct air assault from PZ X (_________) to LZ A (_________) in accordance with Air Movement Table, Appendix 2 of Annex C.

(2) Establish platoon-size blocking positions within OBJ RED (_________) across the major trail leading southwest out of the guerrilla base camp, oriented toward the northeast.

(3) On order, detach one platoon to TF control for air assault reconnaissance missions in selected areas.

(4) Be prepared to assist in the seizure of, or to attack, OBJ WHITE.

c. Co B:

(1) Conduct air assault from PZ Y (_________) to LZ B (_________) in accordance with Air Movement Table, Appendix 2 of Annex C.

(2) Attack and destroy the enemy forces and the guerrilla base camp, OBJ WHITE (_________).

(3) Do not pursue enemy force from OBJ WHITE.

d. Co C:

(1) Conduct air assault from PZ Z (_________) to LZ C (_________) in accordance with Air Movement Table, Appendix 2 of Annex C.

(2) Establish platoon-size blocking positions within OBJ BLUE (_________) across the major trail leading southeast out of the guerrilla base camp, oriented toward the northwest.

(3) Assume attachment of the mortar platoon for movement only.

(4) Be prepared to detach one platoon to TF control for reserve.

e. FSE/1-221 FA: Annex D (Fire Support Overlay).

f. Scout Platoon:

(1) Provide PZ security.

(2) Upon completion of air lift, locate on PZ Y, prepare for contingencymission.

g. Mortar Platoon:

(1) Attached to Co C for movement.

(2) Conduct air assault from PZ Z (_________) to (_________) in accordancewith Air Movement Table, Appendix 2 of Annex C.

(3) Firing position located at (_________), be prepared to displace on order.

(4) Priority of fires initially to Co B, then to Co C.

h. A/21 AHB (OPCON):

(1) Provide air route security to LZ A, B, and C.

(2) Provide supporting fires in priority on LZ B, A, and C from H-2 toH-Hour. On order shift priority to support Co B's attack.

(3) Direct coordination with AMC and Cdr, Co B required.

i. Tm/B/1-11 ACS (OPCON):

(1) Conduct route reconnaissance of all flight routes NLT 110530 Jun.

(2) Provide terminal guidance to LZ A, B, and C for lift assets.

(3) On order, screen to the north of OBJ WHITE.

j. 210 CAC (DS):

(1) Establish laager site vic (_________) from _________ to _________ .

(2) Back haul casualty(s) in event of "hot" LZ.

(3) Execute Air Movement, Annex C (Air Movement Table).

k. 211 CAC (DS):

(1) Establish laager site vic (_________) from _________ to _________ .

(2) Back haul casualty(s) in event of "hot" LZ.

(3) Execute Air Movement, Annex C (Air Movement Table).

l. Coordinating Instructions:

(1) Report LZ status immediately upon landing.

(2) Report all PZs by location and size (in numbers of aircraft) to CP NLT 112100 Jun.

(3) Annex E, OPSCD.

(4) Annex F, OPSEC.

(5) H-Hour is 110530 Jun.

4. SERVICE SUPPORT.

Annex G, Service Support.

5. COMMAND AND SIGNAL.

a. Command.

(1) Command group with Co B for the air assault.

(2) Jump CP to be established with Co B vic (_________).

(3) TOC will displace on order to (_________).

(4) Succession of command: Cdr, XO, S3, Co B Cdr, Co A Cdr.

(5) BN XO will be located in the TOC.

b. Signal.

(1) CEOI #29 in effect.

(2) Air assault net frequency is _________; alternate frequency is _________ .

(a) All stations enter at 110445 Jun.

(b) Ground elements leave net upon securing theLZ.Ground elements send code word BIG SKY when leaving the air assault net.

(c) Assault aviation units remain on assault net until mission complete.

(d) Attack helicopter Co remains on assault net until directed to other nets.

(e) AirCavTeam remains on assault net until directed to other nets by the AATFC.

(3) Emergency signal to lift fires is code word CARDINAL.

(4) Companies mark location with colored smoke as indicated:

Co A - Yellow

Co B - Violet

Co C - Green

(5) Medical evacuation frequency is _________ .

ACKNOWLEDGE.

SMART
LTC

OFFICIAL:

HENRY
S3

ANNEXES:

A - Intelligence (Omitted)
B - Operations Overlay
C - Air Movement
D - Fire Support Overlay
E - OPSCD
F - OPSEC (Omitted)
G - Logistics (Omitted)

DISTRIBUTION: A

Annex B (Operation Overlay) to OPORD 6-85 (Eagle).

Reference: Map, Series B142,

Palomas, Sheet 1011, Special Edition 1, 1:50,000

____
(CLASSIFICATION)

(CLASSIFICATION)

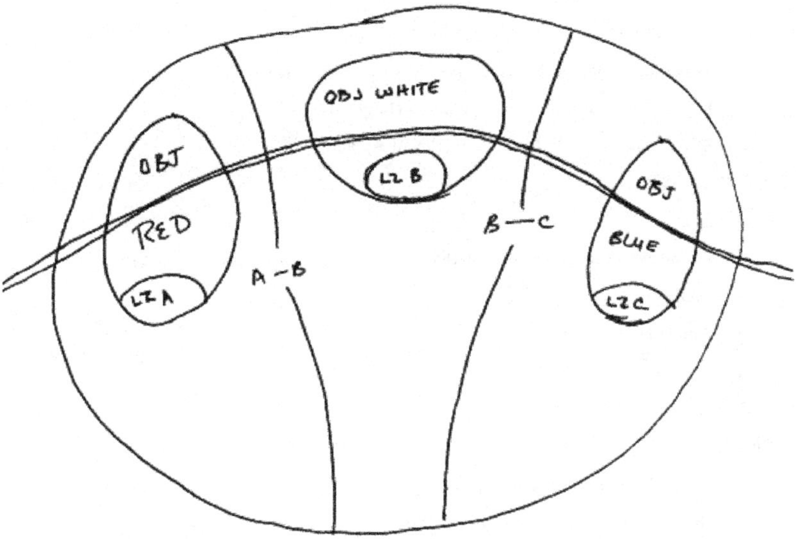

ANNEX C (Air Movement) to OPORD 6-85 (Eagle)

1. SITUATION.

a. Enemy Forces. Annex A, Intelligence.

b. Friendly Forces. Basic OPORD.

c. Attachments and Detachments: Basic OPORD.

2. MISSION.TF1-20 air assaults at 110530 Jun to destroy the enemy forces and the guerrilla base camp, located vicinity coordinates

3. EXECUTION.

a. Concept of Operation. TF 1-20 air assaults from three PZs to multiple LZs via UH-60 with 2 lifts of 24 aircraft each. Companies will assemble and load as shown in Appendix I (PZ Layout). Air assaults will be from designated PZs to designated LZs as per Appendix 2 (Air Movement Table).

b. 210 CAC (DS):

(1)Provide 12 Operational UH-60s.

(2) Provide headset in AMC UH-60 for AATF commander to monitor operation.

c. 211 CAC (DS): Provide 12 operational UH-60s.

d. Coordinating Instructions.

(1) ACL for UH-60 is 12 PAX.

(2) PZmarkedwithhand-heldVS17panelforleadaircraftineachserial.

(3) Flight routes indicated at Appendix 3 (Flight Routes).

(4) SEAD targets on flight route overlay at Appendix 3.

4. SERVICE SUPPORT.

a. Logistics.

(1) Replacement receiving point (RRP) at (_________).

(2) FARP at (_________).

(3) Annex G.

b. Administration. Annex H.

5. COMMAND AND SIGNAL.

a. Signal.

(1) PZ control will be admin-log frequency.

(2) Red colored smoke on LZ means NO LAND - USE ALTERNATE.

b. Command.

(1) AATF commander will ride in AMC aircraft initially.

(2) Basic OPORD.

APPENDIXES:

1 - PZ Layout (Omitted)

2 - Air Movement Table

3 - Flight Routes

_____________________(CLASSIFICATION)

(CLASSIFICATION)

APPENDIX 2 (AIR MOVEMENT TABLE) TO ANNEX C (AIR MOVEMENT) TO OPORD 6-85 (EAGLE)

Copy No ________
Eq ________
Place ________
Date & Time ________
Msg ref No ________

____
(Classification)

Reference: OPORD

Time Zone Used Throughout the Order:

1	2	3	4	5	6	7	8	9	10	11	12	13
AVN UNIT	LIFTED UNIT	LIFT	SERIAL	LOADS	PICKUP ZONE	LOADING TIME	LIFT-OFF TIME	SP TIME	RP TIME	LANDING ZONE	LANDING TIME	REMARKS
210 CAC	A/1-20 (-)	1**	1	1-6	P: X	H-11	H-10	H-8	H-3	P: A A: —	H-Hour	FLT RT - Corr 1 ALTN ____ Bump Load: 6
211 CAC	B/1-20		2	7-17	P: Y A: —	H-11	H-10	H-8	H-3	P: B A: —	H-Hour	FLT RT - Corr 2 ALTN ____ Bump load: 17
210 AVNTM	C/1-20(-)		3	18-24	P: Z A: —	H-11	H-10	H-8	H-3	P: C A: —	H-Hour	FLT RT - Corr 3 ALTN ____ Bump Load: 24
210 CAC	A/1-20 (-)	2***	1	1-5	P: X A: —	H+10	H+11	H+13	H+18	P: A A: —	H+21	FLT RT - Corr 1 ALTN ____ Bump Load: 5
211 CAC	C/1-20 (-)		2	6-9	P: Z A: —	H+10	H+11	H+13	H+18	P: C A: —	H+21	FLT RT - Corr 2 ALTN ____ Bump Load: 9
210 AVNTM	C/1-20 (-)		3	10-13	P: Z A: —	H+10	H+11	H+13	H+18	P: C A: —	H+21	FLT RT - Corr 3 ALTN ____ Bump Load: 13

* 210 AVN TM consists of UH-60s from both the 210 CAC and the 211 CAC

** On Lift 1 211 CAC will receive enough aircraft to lift CO B in a single lift.

*** Only aircraft necessary for insertion will be used. Other aircraft released to laager site to await further instructions.

(CLASSIFICATION)

APPENDIX 3 (FI-IGHT ROUTES) to ANNEX C
(AIR MOVEMENT) to OPORD 6-85 (EAGLE)

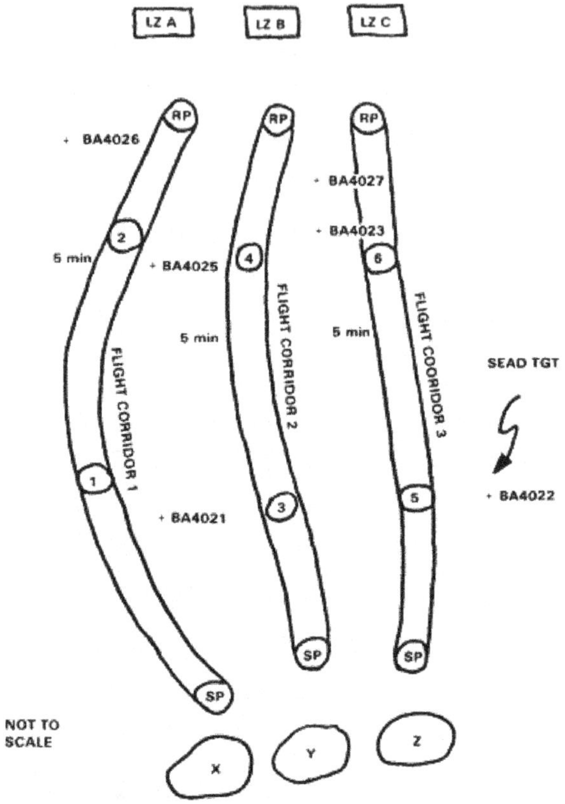

NOTES:

1. SEAD mission based on flak map.
2. LZ prep targets contained in Annex D (Target List).
3. PZ to SP-2 minutes.
4. RP to LZ-3 minutes.

______________________(CLASSIFICATION)

ANNEX D (FIRE SUPPORT OVERLAY) to OPORD 6-85 (EAGLE)

See Appendix 1 to Annex D (Target List). Flight Corridor not to exceed 500 feet (AGL).

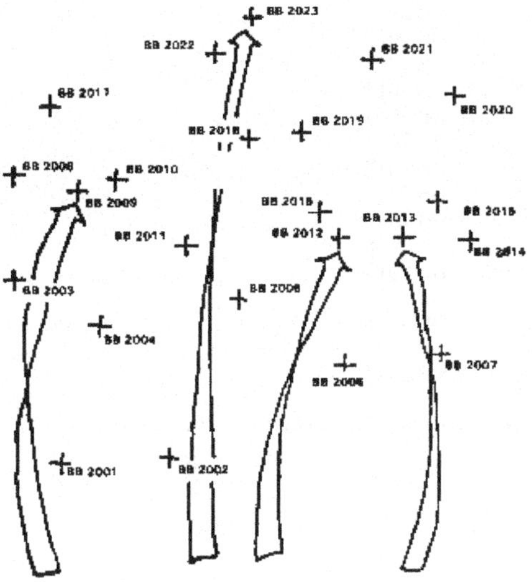

______________________(CLASSIFICATION)

______________________(CLASSIFICATION)

APPENDIX 1 (TARGET LIST) to ANNEX D (FIRE SUPPORT OVERLAY)
to OPORD 6-85 (EAGLE)

LINE NUMBER	TARGET NUMBER	GRID	DESCRIPTION

1	BB 2001	(Left blank intentionally)
2	BB 2002	
3	BB 2003	
4	BB 2004	
5	BB 2005	
6	BB 2006	
7	BB 2007	
8	BB 2008	
9	BB 2009	
10	BB 2010	
11	BB 2011	
12	BB 2012	
13	BB 2013	
14	BB 2014	
15	BB 2015	
16	BB 2016	
17	BB 2017	
18	BB 2018	
19	BB 2019	
20	BB 2020	
21	BB 2021	
22	BB 2022	
23	BB 2023	

Target Grid and description based upon support of current tactical plan.

______________________(CLASSIFICATION)

______________________(CLASSIFICATION)

ANNEX E (OPSCD) to OPORD 6-85 (EAGLE)

NUMBER	**TIME**	**UNIT**	**ACTIVITY**
1	H-60	TM/6/1-11 ACS	Commence route reconnaissance with FAC aboard.
2	H-30	AATF	PZ posture on all PZs.
2a		TM/B/1-11	ACS Report progress on route reconnaissance.
3	H-20	TM/B/1-11	ACS Reconnaissance LZs A, B, C.
4	H-15	TM/B/1-11 ACS	Report LZ status of LZs A, B, C.
5	H-12	A/21 AHB	Rendezvous at LZs X, Y, Z. Provide en route security.
6		AATF	Load aircraft.
7	H-10	AATF	Depart PZs.
7a		A/1-20	Depart PZ X.

7b		B/1-20	Depart PZ Y.
7c		C/1-20	Depart PZ Z.
8	H-5	1-221 FA	Commence prep fires on LZs A, B, C.
9	H-3	AATF	En route from RP.
9a		TM/B/1-11 ACS	Rendezvous with lead elements of AATF for terminal guidance.
10	H-2	1-221 FA	Cease fire on prep.
10a		A/21 AHB	Commence 2-min prep fire on LZs A, B, C.
11	H-Hour	AATF	Land at LZs A, B, C.
11a		A/1-20	Land at LZ A.
11b		B/1-20	Land at LZ B.
11c		C/1-20	Land at LZ C.
11d		TM/A/21 AHB	Overwatch LZs A, B, C.
12	H-Hour	210 CAC, 211 CAC	En route to PZs X, Y, Z.
12a		TM/A/21 AHB	En route, air assault route security.
13	H+10	210 CAC 211 CAC 210 AVN TM	Arrive at PZs X, Y, Z.
13a		1-20	Load aircraft.
14	H+11	210 CAC 211 CAC 210 AVN TM	Depart LZs X, Y, Z.
14a		A/1-20	Depart LZ X.
14b		B/1-20	Depart LZ Y.
14c		C/1-20	Depart LZ Z.
14d		TM/A/21 AHB	En route, air assault route security.
15	H+18	210 CAC 211 CAC 210 AVN TM	En route from RP.
15a		1-20	En route from RP.
15b		TM/B/1-11 ACS	Rendezvous with lead elements of AATF for terminal guidance.
15c		TM/A/21 AHB	En route, air assault route security.
16	H+21	210 CAC 211 CAC 210 AVN TM	Land at LZs A, B, C.
16a		A/1-20	Land at LZ A.
16b		A/1-20	Land at LZ B.
16c		A/1-20	Land at LZ C.
17	H+22	AATF	Completion of air assault phase.

NOTE: OPSCD can be carried out throughout the entire operation.

______________________(CLASSIFICATION)

APPENDIX B

Air Assault Planning Forms

Section I

PREPARATION OF THE AIR MOVEMENT TABLE

B-1. General.

The chart illustrated in this section is a planning guide (Chart B-1).

B-2. Column 1.

The aviation unit commander will enter the unit designation (by subelement) in this column (for example, A/121CAB designates Co A, 121st Combat Aviation Brigade).

B-3. Column 2.

The ground unit's designation (by subelement) will be entered in this column in its order of commitment (for example, A/1-66 designates Co A, lst Bn, 66th Infantry).

B-4. Column 3.

Enter lift numbers, beginning with No. 1. When all available aircraft have been listed (see column 5 with 36 aircraft available), then lift No. 2 and so on.

NOTE: Elements of one or more units may be moved in the same lift.

B-5. Column 4.

a. Serials within each lift will be numbered sequentially, beginning with No. 1. On lift No. 2, the serial numbers will again start with No. 1.

b. Serials are created by subdividing lifts. This is accomplished to:

(1) Reduce span of aircraft control (no more than 6 aircraft per serial).

(2) Maintain control when multiple pickup zones, landing zones, and/or flight routes are used.

(3) Maintain control when the elements of the lift are separated.

B-6. Column 5.

Loads are numbered sequentially, one for each aircraft allocated, within a lift. The number of loads per serial is dependent upon the number of PZs, LZs, and flight routes to be used in the operation.

B-7. Column 6.

Pickup zones should be listed as primary (prim) and alternate (altn) for each serial. This provides continuity when naming PZs. If they are named blue, green, amber, and white, it facilitates their marking by pathfinders (with the colored light sets available for night operations and smoke or panels for day operations). Red is used to mark obstacles. Formations desired in a PZ should be listed under the alternate PZ code name (for example, staggered trail right [S/R], staggered trail left [S/L]). For refueling, list the FARP designation where the aircraft are to be refueled.

B-8. Column 7.

a. List loading times and/or refueling times.

b. Loading time is the time at which troops and equipment begin loading into the aircraft. Load time will be specified by the PZ control officer. When FARP numbers are listed in column 6, and "refuel" is listed in column 8, the time in column 7 indicates the time aircraft are scheduled for refueling.

B-9. Column 8.

The lift-off time is the time an aircraft departs the PZ. This time should be planned to allow aircraft to depart the PZ, assemble into their in-flight formation, and arrive at the start point at the required time. Normally, 2 or 3 minutes prior to SP time is sufficient. When refueling is entered in this block and subsequent blocks, the aircraft is scheduled to refuel.

B-10. Column 9.

List the time aircraft are scheduled to cross the SP. It normally takes 2 to 3 minutes to fly from the PZ to the SP (SP 3 to 5 km from PZ). To arrive at a realistic SP time, determine the flight time between the SP and release point. Subtract this time from the RP time, and the result is the SP time.

a. List the time aircraft are scheduled to cross the RP. To derive the time required to fly from the SP to the RP, measure the distance of the route (on a map), obtain the average ground speed from the AMC, and compute.

b. The RP time is normally 2 to 3 minutes or 3 to 5 km from the LZ, which facilitates closely coordinated fire support control.

B-12. Column 11.

List primary and alternate LZs. They will be designated for each serial. Unit SOPs dictate the method of marking LZs. The desired landing formation will be listed under the LZ code name (for example, S/R, S/L). PZ and LZ names must not conflict.

B-13. Column 12.

List the landing time for the first aircraft of each serial. All times are listed in terms of H-hour (time the first aircraft of the first serial touches down in the LZ). All times in columns 7, 8, 9, and 10 are listed with reference to H-hour, either minus (time prior to H-hour) to plus (time after H-hour).

B-14. Column 13.

Remarks will contain any information necessary to ensure that the air movement table is understood; for example, flight altitude, flight routes (prim, altn, rtn, and sequence of bump by load). (Bump 17 and 14 means that load 17 would be bumped first.)

Section II

PREPARATION OF THE AIRLOADING TABLE

B- 15. General.

The chart illustrated in this section is a planning guide (Chart B-2).

Reference: Map OPORD

Classification

P. Primary
A. Alternate

Time Zone Used Throughout the Order: ALFA

Copy No ____________
Hq ____________
Place ____________
Date & Time ____________
Msg Ref No ____________

1	2	3	4	5	6	7	8	9	10	11	12	13
AVN UNIT	LIFTED UNIT	LIFT	SERIAL	LOADS	PICKUP ZONE	LOADING TIME	LIFT OFF TIME	SP TIME	RP TIME	LANDING ZONE	LANDING TIME	REMARKS
A/121 CAB	A/1 66	1	1	1 6	P COW	H-19	H-17	H-15	H-2	1,A-1A	H	P FLT RT SIERRA
					A COW 1					S L		A XRAY
			2	7 12		H-18	H-16	H-14	H-1		H-1	ROMEO (RTN)
			3	13 18	S L	H-17	H-15	H-13	H		H-2	BUMP 17 14
B 121 CAB	B/1 66		4	19 27	P DOG	H-19	H-17	H-15	H-2	2,A-3	H	P FLT RT ZULU
					A DOG 1					S L		A-OSCAR (RTN)
			5	28 36		H-18	H-16	H-14	H-1		H-1	
					S L							BUMP 21 29
A 121 CAB	C 1 66	2	1	1 9	P CAT	H-18	H-20	H-22	H-35	5,A-4	H-37	P FLT RT SIERRA
					P CAT 1					S R		A XRAY
			2	10 18		H-19	H-21	H-23	H-36		H-38	ROMEO (RTN)
					S L							BUMP 7 17
B 121 CAB			3	19 27		H-20	H-22	H-24	H-37		H-39	
			4	28 36		H-21	H-23	H-25	H-38		H-40	
A 121 CAB	C 1 66	3	1	1 9	FARP #1	H-55	REFUEL					
				10 18	FARP #2	H-56	REFUEL					

Chart B-1. Air Movement table.

B-16. Column 1.

The information in this column is provided by the squad leader to the platoon leader (after the squad leader has been given the allowable cargo load). The squad leader(s) will normally write this information on sheets of notebook paper which will be assembled by the platoon leader. This list of personnel and equipment can also serve as the aircraft manifest. Also, it may indicate a sequence of bump for the personnel on board if a pilot indicates that he cannot safely lift the planned ACL. The bump sequence will start at the bottom of the list and go upward. That means that the person listed last on the list would be bumped first. If two people had to be bumped, the last and next to last person listed would be bumped.

B-17. Column 2.

At battalion level, the PZ, altn PZ, and PZ formation will normally be determined by the battalion pickup zone control officer. At company level, this information will be taken from the battalion air movement table.

B-18. Column 3.

The PZ arrival time of the unit will be designated by the PZCO. This is the time when the troops arrive at the PZ to prepare aircraft loads adjacent to proposed touchdown points for the helicopters.

B-19. Column 4.

This information will be taken from the battalion air movement table. It *is* the time that troops begin loading on the aircraft.

B-20. Column 5.

This information will be taken from the battalion air movement table. It will provide the sequence of movement within the battalion and company.

B-21. Columns 6 and 7.

This information will be taken from the battalion air movement table. It will indicate which aviation unit will be lifting a particular load.

B-22. Column 8.

Each aircraft within the lift is assigned a load number. The numbers will correspond to a specific lift and serial; for example, lift No. 1, serial 2, load 7. The platoon will be given specific load numbers by the company commander, and the platoon leader will in turn assign a load number to each of his aircraft loads. This information will normally not be available at the time the squad leader prepares his manifest, but should be provided to him as soon as possible.

B-23. Column 9.

The remarks column can be used by the commander to place specific information not provided in the preceding columns. One pertinent bit of information that is essential in this column is the sequence of bump for aircraft loads; for example, serial 1, loads 4-5; serial 2, loads 9-10. This means that if two loads cannot be lifted, loads 4 and 5 will be the ones that remain in the PZ. If only one load cannot be lifted, load 4 will remain in the PZ.

Reference: OPORD (Classification)

Time Zone Used Throughout the Order: ALFA

Copy No ________
Hq ________
Place ________
Date & Time ________
Msg ref No ________

1	2	3	4	5	6	7	8	9
Personnel & Equipment	PZ	PZ Arrival Time	Loading Time	Avn Unit	Lift No.	Serial No.	Load No.	Remarks
1-1-A-1/66 INF 1-1A SSG JONES 1-1A SGT SELECIA 1-1A SP4 MELINDEN 1-1-A SP 4 BAIER 1-1-A PFC NANNI 1-1-A PFC BYRD 1-1-A PVT JOHNSON 1-1-A	BLUE ALTN BLUE 1 S/L	H-34	H-19	A/121CAB	1	1	1	SEQUENCE OF BUMP. SERIAL 1 LOADS 4 & 5
1 & 2-A-1/66 INF SSG SLOAN 2-1-A SGT ALVIS 1-1-A SP4 WILLIAMS 1-1-A PFC YOUNG 1-1-A PFC GREEN 2-1-A PVT BARAJAS 1-1-A SGT ULMER 2-1-A	BLUE ALTN BLUE 1 S/L	H-34	H-19	A/121CAB	1	1	2	SEQUENCE OF BUMP. SERIAL 2, LOADS 9 & 10

Chart B-2. Airloading table.

Section III

SEQUENCE OF ACTIONS FOR AIR ASSAULT OPERATIONS

B-24. General.

The following lists provide a sequence of actions taken by the air assault task force commander, his staff, and subordinate unit commanders in planning an air assault operation. The list is not all inclusive. Certain actions may be omitted for some operations.

B-25. Air assault task force commander actions.

a. Receives brigade warning order.

b. Conducts mission analysis.

c. Receives air mission commander's initial information.

d. Gives warning order to task force.

e. Receives personnel status report from S1.

f. Receives equipment status report from S4.

g. Receives enemy situation briefing from S2.

h. Receives friendly forces information briefing from S3.

i. Continues analysis of mission, enemy, terrain (and weather), troops, and time (available).

j. Receives brigade OPORD.

k. Begins development of commander's estimate.

l. Provides guidance to staff as needed.

m. Receives staff estimates.

n. Obtains data from staff as needed.

o. Announces concept.

p. Supervises development of OPORD.

q. Receives air movement table. Approves, or modifies and approves.

r. Receives airloading tables from S3.

s. Receives copy of OPORD from S3.

t. Approves, or modifies and approves, OPORD. Returns OPORD to S3.

u. Issues, or oversees issuance of, OPORD.

B-26. Air mission commander actions.

a. Receives brigade warning order.

b. Conducts mission analysis.

c. Gives AMC initial information to AATF commander and staff.

d. Receives task force warning order.

e. Receives friendly forces information briefing from AATF S3.

f. Provides technical advice to AATF executive officer and S2 for PZ and LZ identification.

g. Coordinates with supported unit staff.

h. Provides information to aviation unit on ground unit operation.

i. Provides advice to AATF S3 on PZ selection.

j. Provides flight route information to AATF S2.

k. Assists AATF XO in PZ control plan.

l. Assists AATF S3 Air with flight route computations.

m. Coordinates PZ operations with AATF XO.

n. Provides advice to AATF S3 on LZ and flight route selection.

o. Obtains PZ and/or LZ, flight routes, and aircraft allocation from AATF S3 Air.

p. Obtains PZ control plan from AATF XO.

q. Aids AATF S4 in selecting logistic PZ(s).

r. Coordinates aircraft internal and sling equipment loads with AATF S4.

s. Obtains air movement table for AATF S3 Air.

t. Briefs aviation unit on operation.

u. Inspects PZ(s) with AATF XO.

v. Receives AATF OPORD.

B-27. Executive officer actions.

a. Receives TF warning order.

b. Receives personnel status report from S1.

c. Receives equipment status report from S4.

d. Receives enemy situation briefing from S2.

e. Receives AMC initial information.

f. Receives friendly forces information briefing from S3.

g. Determines available PZs. Obtains advice from AMC.

h. Submits PZs to S3.

i. Coordinates staff planning.

j. Obtains PZs from S3.

k. Develops PZ control plan.

l. Coordinates PZ operations with AMC and/or pathfinders.

m. Receives AATF commander's concept.

n. Obtains PZs, LZs, flight routes, and aircraft allocation from S3 Air.

o. Coordinates PZ operation with Sl. Completes bump and straggler control plan.

p. Inspects PZs with pathfinders.

q. Briefs PZ control plan to subordinate PZCOS. Provides pathfinder to each PZCO.

r. Obtains air movement table from S3 Air.

s. Obtains airloading tables from S3 Air.

t. Obtains sequence of bump from subordinate units. Annotates air movement table with sequence of bump.

u. Completes PZ control plan. Submits to S3.

v. Inspects PZs with pathfinder and/or AMC.

w. Receives OPORD.

B-28. S1 actions.

a. Receives operation notification.

b. Assembles personnel data.

c. Receives task force warning order.

d. Reports personnel status to commander and staff.

e. Receives AMC initial information.

f. Receives friendly forces information briefing from S3.

g.Begins mission analysis from personnel standpoint.

h. Begins preparation of staff appraisal (personnel).

i. Completes health service support plans.

j. Provides S3 with staff appraisal (personnel).

k. Receives AATF commander's concept.

l. Coordinates PZ operations with battalion XO. Develops straggler control plan.

m. Briefs subordinate unit personnel on straggler control plan.

n. Receives command post's general location from S3.

o. Coordinates prisoner of war (PW), civilian control plan with S2.

p. Completes PW, civilian control plan. Coordinates with S4.

q. Completes S1 portion of paragraph 4, OPORD. Gives to S4.

r. Receives airloading tables from S3 Air.

s. Coordinates with headquarters (HQ) commandant. Develops plan for command post displacement and security.

t. Coordinates command post displacement plan with S3.

u. Plans for recovery and evacuation of the dead, coordinates with S3 and S4.

v. Receives OPORD.

B-29. S2 actions.

a. Receives operation notification.

b. Requests any needed maps through S4.

c. Assembles intelligence data.

d. Requests weather forecast,

e. Distributes maps.

f. Briefs commander, staff, and subordinate unit commanders on enemy situation.

g. Obtains advice concerning LZs from AMC and pathfinders.

h. Determines available LZs.

i. Submits LZ list to S3.

j. Analyzes weather forecast.

k. Obtains advice concerning flight routes from AMC.

l. Determines available flight routes.

m. Provides available flight routes to S3 Air.

n. Recommends priority intelligence requirements and information requirements.

o. Develops intelligence collection plan.

p. Requests aviation reconnaissance and/or aerial photographs.

q. Begins preparation of staff appraisal (intelligence).

r. Provides intelligence data to commanders and staffs.

s. Processes intelligence data gathered.

t. Completes staff appraisal (intelligence).

u. Provides S3 with staff appraisal (intelligence).

v. Continues processing intelligence data; provides to commanders and staffs.

w. Recommends employment of ground surveillance radar or scout platoon to S3.

x. Recommends targets to FSCOORD.

y. Completes paragraph I (enemy forces) of OPORD and submits to S3.

z. Continues processing intelligence data; provides to commanders and staffs.

aa. Coordinates development of PW and civilian control plan with SI.

bb. Provides all intelligence data to commanders and staffs as it becomes available (aerial photographs, reconnaissance mission results).

cc. Updates intelligence map as needed.

dd. Continues processing intelligence data. Provides to commanders and staffs.

ee. Receives OPORD.

B-30. S3 actions.

a. Receives brigade warning order.

b. Assembles data on friendly elements.

c. Receives AMC initial information.

d. Receives TF warning order.

e. Receives personnel status from S1.

f. Receives equipment status from S4.

g. Receives enemy situation briefing from S2.

h. Briefs AMC initial information.

i. Briefs friendly forces disposition and location.

j. Begins development of courses of action.

k. Obtains input on pathfinder employment.

l. Receives list of available LZs from S2; available PZs from XO.

m. Receives brigade OPORD.

n. Begins preparation of staff appraisal (operations).

o. Selects PZs. Briefs XO and S3 Air on PZs selected.

p. Obtains available flight routes from S3 Air.

q. Consolidates staff information.

r. Selects LZs and flight routes.

s. Determines need for indirect fire preparations.

t. Determines need for EW support.

u. Provides staff appraisal to commander.

v. Receives commander's decision.

w. Begins preparation of OPORD.

x. Receives employment recommendations for ground surveillance radar (GSR) and scout platoon from S2.

y. Provides XO and S3 Air with PZs and aircraft allocation.

z. Selects general location for command post. Provides information to Sl.

aa. Receives S2 input to OPORD. Receives administrative-logistical (admin-log) portion of order from S4.

bb. Completes OPORD paragraph 1, 2, and 3 (to include concept and subunit paragraph).

cc. Receives air movement table from S3 Air. Obtains approval from AATF.

dd. Receives fire plan from FSCOORD.

ee. Receives EW support plan from MI support unit or liaison personnel.

ff. Obtains airloading tables from S3 Air.

gg. Receives OPORD paragraph 4 information from S4.

hh. Completes paragraph 5, OPORD.

ii. Completes operation overlay.

jj. Coordinates command post displacement with SI.

kk. Obtains PZ control plan from XO.

ll. Completes OPORD with annexes. Submits to AATF commander for approval.

mm. Receives OPORD from AATF commander.

nn. Issues OPORD, when directed by commander.

B-31. S3 Air actions.

a. Receives operation notification.

b. Receives TF warning order.

c. Receives personnel status from SI.

d. Receives equipment status from S4.

e. Receives enemy situation briefing from S2.

f. Receives AMC initial information.

g. Analyzes AMC's initial information for available assets.

h. Begins preparation of air movement table.

i. Processes air requests from S3.

j . Obtains PZs from S3; provides PZs to fire support coordinator and staff as needed.

k. Establishes liaison with TACAIR control party; coordinates preplanned TACAIR.

1. Receives available flight routes from S2.

m. Computes flight route time and distance.

n. Provides available flight route information to S3.

o. Receives AATF commander's concept.

p. Obtains LZs, flight routes, and aircraft allocation from S3. Provides data to AMC, FSCOORD, subordinate unit commanders, and staff, as needed.

q. Obtains any additional tactical air requirements from FSCOORD.

r. Initiates requests for TACAIR cap (cover) and offensive air support.

s. Obtains logistic PZs from S4.

t. Completes air movement table; submits to S3 for approval.

u. Distributes air movement table.

v. Obtains subunit airloading tables.

w. Consolidates airloading tables; provides to AATF commander, S3, XO, and Si.

x. Receives OPORD.

B-32. S4 actions.

a. Receives operation notification.

h. Obtains maps requested by S2.

c. Assembles equipment data.

d. Receives TF warning order.

e. Receives personnel status from SI.

f. Reports equipment status to commander and staff.

g. Receives enemy situation briefing from S2.

h. Receives AMC's initial information.

i. Receives friendly forces information from S3.

j. Begins mission analysis to determine requirements.

k. Receives initial supply requirements from subordinate units.

l. Begins preparation of staff appraisal (logistics).

m. Determines effects of ammunition supply rate on operation. Submits ammunition supply rate to FSCOORD.

n. Compiles materiel usage data for operation. Obtains PZs from S3 Air.

o. Compares usage data to materiel available.

p. Requests materiel as needed.

q. Coordinates with AMC on establishing FARP.

r. Provides S3 with staff appraisal (logistics).

s. Receives AATF commander's concept.

t. Begins development of support plan for operation. Obtains LZs and flight routes from S3 Air.

u. Coordinates trains operation and control with supply platoon leader. Obtains Sl input to paragraph 4, OPORD.

v. Completes S4 portion of paragraph 4, OPORD. Provides paragraph 4, OPORD, information to S3.

w. Selects logistics PZs; provides to S3 Air.

x. Plans aircraft loads (internal and sling) for mission support. Coordinates pickup points with AMC and/or pathfinder.

y. Coordinates plans for evacuation of enemy materiel with S2.

z. Receives OPORD.

B-33. FSCOORD actions.

a. Receives operation notification.

b. Begins mission analysis to determine available and needed means of fire support.

c. Plots locations and capabilities (range fans) for all indirect fire support systems supporting the TF.

d. Estimates fire support needed.

e. Obtains ammunition supply rate from TF S4 and S3 of DS FA battalion. Determines effects of ammunition supply rate on operation. Gathers information for development of fire support plan.

f. Obtains PZs from S3 Air.

g. Continues to gather information for development of fire support plan.

h. Coordinates fire support requirements with S3.

i. Provides S3 with available fire support recommendation for indirect fire preparations.

j. Obtains LZs and flight routes from S3 Air.

k. Obtains recommended targets from S2. Develops TACAIR requests to support ground tactical plan.

l. Completes fire support plan.

m. Submits fire support plan to S3 for commander's approval; on approval, distributes.

NOTE: Subunits will develop their fire plans. FSCOORD will coordinate and consolidate them into the AATF plan.

n. Receives OPORD.

B-34. Subordinate unit commanders' actions.

a. Receives operation notification.

b. Gathers personnel and equipment data.

c. Reports personnel and equipment to battalion staff. Receives maps.

d. Receives battalion warning order.

e. Issues company warning order.

f. Determines initial supply requirements.

g. Submits initial supply requirements to S4.

h. Begins preparation of airloading tables.

i. Continues mission preparation.

j. Obtains appropriate PZs, LZs, flight routes, and aircraft allocation from S3 Air.

k. Continues mission planning.

l. Obtains air movement table from S3 Air.

m. Completes airloading tables.

n. Submits airloading tables to S3 Air.

o. Receives OPORD.

APPENDIX C

Sample Annex to Infantry Battalion SOP for Air Assault Task Force Operations

With enclosures:

1. Air Assault Formations and Battle Drill.

2. Air Assault Task Force Commander's Checklist.

3. Air Mission Commander's, or Air Liaison Officer's, Checklist.

NOTE: This is an annex to a standing operating procedure for combat operations; it is not a complete SOP.

______________________(CLASSIFICATION)

Copy__of__ Copies
1st Battalion, 66th Infantry
Fort Benning, Georgia
25 December 19__

Annex E (Air Assault Operations) to _____ Infantry

Battalion, _____ Infantry Division Field Standing

Operating Procedures

1. GENERAL.

a. Purpose. This annex prescribes the organization and procedures to be followed in preparing for and executing air assault task force operations. Only procedures peculiar to air assault operations are included; otherwise, basic standing operating procedures apply.

b. Application. Applies to all organic and supporting units under control of 1st Battalion, 66th Infantry. Company SOPs will conform.

2. PERSONNEL.

a. Strength, Records, and Reports.

(1) Companies will be organized into assault and rear echelons. Upon receipt of warning

order, submit unit strength to S1 and equipment availability status to S4. S1 and S4 will forward to S3 who will determine sortie requirements.

(2) Upon closing into landing zone, companies will report personnel and equipment status to AATF forward command post on AATF command net using report format in CEOI.

b. Discipline, Law, and Order.

(1) S1 will establish a straggler control point on each pickup zone in vicinity of pickup zone control officer in coordination with S3 and the PZCO. All units will have a representative (from rear echelon) located at the straggler control point on their PZ(s). "Bumped" personnel will be reported to S1 and/or PZCO by company straggler control personnel for consolidation and rescheduling into appropriate LZ(s).

(2) Straggler control is company responsibility upon landing.

(3) Personnel landed in other than an assigned LZ are to report to the on-site unit commander (representative) immediately. Personnel are attached to that unit until ordered to return to parent unit by this headquarters. Gaining unit will report personnel so attached to S1 by number and parent unit. (Do not include these personnel in unit strength reports.)

c. Enemy Prisoners of War (EPW). EPWs are to be reported immediately to the S3, who will issue evacuation instructions. Indicate available PZ location for pickup by air in initial report. S2 will determine whether to evacuate through battalion or direct to brigade PW collection point.

d. Medical Evacuation. Report all casualties for medevac by priority code.

(1) Medevac requests follow standard format for casualties and are classified as routine, priority, or urgent. Radio frequency of supporting medevac unit will be included in each operation order. When medevac aircraft are not available and immediate medevac is required, make maximum use of empty lift helicopters departing LZ. **Do not interrupt airlanding operations** - use last helicopters in the flight.

(2) S1 is responsible for providing CEOI, LZs, and flight route overlay to supporting medevac unit.

3. INTELLIGENCE.

a. Weather.

(1) Battalion S2 will obtain and disseminate the following:

(a) Long-range forecast immediately after receipt of mission.

(b) Short-range forecasts up to H-2 hours.

(2) Command weather reconnaissance 1 hour prior to lift-off will be coordinated among task force commander, S2, and air mission commander.

(3) Operations are executed only on order of this headquarters when weather is below 1/2-mile visibility and 100-foot ceiling.

b. Terrain.

(1) Maximum utilization will be made of command aerial reconnaissance down to platoon leaders, consistent with aviation resources, available time, and tactical situation.

(2) Use sandtable briefing techniques when possible in conjunction with maps and aerial photographs.

(3) Maps will be issued immediately after receipt of warning order, if available. If not available, they will be issued on receipt from higher headquarters.

(4) Aerial photographs will be made available upon receipt. S2 will automatically process all requests (conserving assets, as appropriate). Priority to answering essential elements of information (EEI). Priority of issue to assault echelon.

c. Evasion and Escape.

(1) Personnel in aircraft forced to land behind enemy lines en route to the objective will:

(a) Establish immediate security in vicinity of downed aircraft.

(b) Remain in that location utilizing aircraft radios to contact recovery aircraft.

(c) Mark and clear suitable landing points for recovery and withdrawal helicopters.

(2) The senior ground force individual assumes responsibility for organization and conduct of security until recovery is executed.

(3) If the above is not possible due to enemy pressure, evade capture and attempt to join friendly units by infiltration. Personnel will attempt to evade and escape back along flight route. Maximum rescue effort will be directed along flight routes, with particular emphasis at checkpoints. Continuous attempts will be made to locate suitable landing points for withdrawal by helicopter(s). The wounded will be evacuated with infiltrating personnel. The dead will be concealed and stripped of weapons, ammunition, and items of intelligence value. Locations of dead left behind will be recorded,

(4) If enemy pressure becomes a threat to downed aircraft personnel, senior ground force individual will take steps to secure or destroy classified or sensitive items. Aircraft destruction will be on order of this headquarters if capture is not imminent. If contact with this headquarters cannot be made, the senior individual on the ground will make decision on aircraft destruction in order to prevent capture by enemy.

4. OPERATIONS.

a. Planning Phase.

(1) Except when accomplished by higher headquarters, this headquarters will prepare plans in coordination with the supporting AMC.

(2) Plans will continue to be refined until executed. All operational information will be given to subordinate commanders as soon as determined, particularly as follows:

(a) The size and composition of the force required to execute the mission.

(b) Allocation of assault and logistical aircraft, based on allowable cargo load provided by the AMC.

(c) Designated PZs and helicopter PZ formation. Designated flight routes, LZs, and LZ helicopter formation.

(d) Current communications-electronics operation instructions in effect to include frequencies and call signs of all participating units.

(3) Coordination between the supported and supporting commanders must include, as a minimum:

(a) Enemy and friendly situation.

(b) Mission.

(c) Fire support plan.

(d) Abort and alternate plans.

(e) Weather to include minimums and delays.

(f) Type, number, and ACL of helicopters.

(g) Helicopter formations in PZ and LZ.

(h) Air movement data and timing for the operations.

(i) Communication (primary and alternate frequencies and plans).

(j) Location and call sign of second in command.

(k) Required command reconnaissance by the air assault TF and supporting aviation commanders.

(l) Time synchronization.

(m) Downed aircraft procedures.

(4) Operations security. This will be emphasized in each phase of an air assault operation. The object will be to conceal the capabilities and intentions of the AATF. The four general OPSEC measures will be considered for every operation: deception, signal security, physical security, information security. The S2 will provide intelligence collection of Threat data. The S3 will ensure that the staff and subordinate commanders are aware of OPSEC measures to be employed to counter the Threat. Emphasis must be placed on maintaining the elements of surprise and security. Additionally, all supporting elements must be aware of the necessity of maintaining a high degree of operational security. As a minimum, the AATF commander, supporting commanders, and subordinate commanders should employ the following techniques:

(a) Deception.

- Camouflage vehicles, equipment, and personnel.
- Overflights of other LZs (if aircraft are available and enemy situation does not preclude).
- Insertion at night or during other periods of reduced visibility.
- Noise and light discipline.
- Dummy laager sites for aircraft.
- Recon overflights of several objectives.

(b) Signal security (SIGSEC).

- Radio Communications security techniques.
- listening silence.
- Use of arm-and-hand signals (on the ground).
- Use of low power and secure mode on radios.

(c) Physical security.

- Use of security forces at LZ and PZ.
- Use of wire, mines, barriers, and security troops at aircraft laager sites and troop assembly areas.
- Use of pathfinder to secure LZ, if possible.

(d) Information security.

- Counterintelligence.
- Strict control of all operational information.
- Release information only to those with a need to know.
- Last-minute release of attack time (objective) and force composition.

b. Landing Phase.

(1) The aircraft commander will notify each aircraft troop commander of any changes to the order, any change in LZ(s) and/or direction of landing, and when the helicopter is over the release point. The troop commander then informs his personnel of any changes and alerts them to prepare to unload.

(2) Passengers may not move in the aircraft until clearance has been obtained from the aircraft commander. After the aircraft commander gives the clearance signal, troops and equipment are unloaded as rapidly as possible.

(3) After all troops and cargo have been unloaded from the aircraft, the crew chief will check the helicopter and signal the aircraft commander that the cabin is empty. Personnel will not depart helicopters to the rear. Departure from aircraft will be executed rapidly in the direction prescribed by battle drill (see battle drill, enclosure 1).

(4) The troop commander ensures that members of his aircraft clear the LZ in a safe, expeditious manner. This prevents exposing personnel to unnecessary danger and prevents any delay in lift-off and landing of subsequent helicopters.

(5) Individual weapons will be fired only on order upon offloading unless enemy contact is made or if planned as part of the overall fire plan.

(6) Actions when there is no enemy contact on the LZ include:

(a) Move each helicopter load to the nearest covered and concealed position in direction of assembly area.

(b) Establish LZ security for succeeding lifts (if applicable).

(c) Assemble, organize, and account for all personnel.

(d) Report.

(7) Actions when enemy contact is made on the LZ include:

(a) Return fire immediately, upon offloading, with all available firepower to gain fire superiority.

(b) Fight by helicopter loads, using fire and movement, until platoon or company can be formed (see battle drill, enclosure 1).

(c) Request and coordinate fire support.

(d) Secure the LZ for succeeding lifts.

(e) Report.

(8) Keep the AATF commander informed during all actions.

c. Air Movement Phase.

(1) Lift-off time; passing start points, RP(s), communication checkpoints, and LZ clearance time; and situation on LZ are reported to command and control helicopter by each aviation serial commander. (This may be omitted if operation order specifies radio silence.) Inability to comply with specific control times will also be reported as prescribed in OPORD.

(2) Troop leader remains oriented by continuous map-terrain comparisons.

d. Loading Plan.

(1) PZ is designated by this headquarters.

(2) Aviation serial commander or aviation liaison officer will arrive prior to the helicopter flight and report to PZCO for last-minute briefing and coordination. He will notify the AMC of any changes and (along with pathfinders) will control aircraft operations.

(3) Serials organized to support the ground tactical plan.

(4) Helicopters will land in the PZ(s) in the specified formation (see enclosure 1). Unit leaders will brief troops on the helicopter formation prior to arrival of helicopters at PZ. PZs will be designated and marked using standard North Atlantic Treaty Organization (NATO) colors (PZ Green marked with green smoke or helmet liner during the day, and green lights

at night unless this will compromise security).

(5) Helicopters should arrive at the latest possible time to reduce time-on-ground before loading.

(6) During a battalion move, the battalion executive officer, or designated representative, will act as PZCO. Company XO will act as PZCO during company-size operations and as unit PZCO during battalion-size operations with separate company PZ. Platoon sergeants will act as PZCO during platoon-size operations. PZCOs are responsible for developing and disseminating the PZ control plan. The PZCO will maintain contact with the AMC on a designated radio frequency personally or through assigned pathfinders. Each unit to be moved will have radio contact with the PZCO on a designated frequency. These personnel will establish radio contact with the PZCO 15 minutes prior to aircraft arrival. Units must be prepared to alter loads based on change of helicopter availability or change in allowable cargo load. Within each company, platoon, and squad, a priority of loading must be established. Priority of aircraft loads and personnel on each aircraft to be bumped will be designated. Personnel bumped report to the straggler control point.

(7) Supporting aviation unit assists in planning for the execution of loading by providing technical advice, supervision, and pathfinder support.

(8) Aircraft commander supervises aircraft loading.

(9) Cargo or equipment to be transported externally is secured in cargo nets or on pallets for slingloading under helicopters. Hookup of these loads will be accomplished by the PZ control group. Unit code panels will be placed on each load.

(10) The following individual preparation will be accomplished:

(a) Fasten helmet chinstraps.

(b) Collapse bipods on M60s and M16s.

(c) Tie down loose equipment.

(d) Unload all weapons and place them on SAFE.

(e) Unfix bayonets (if fixed).

(f) Wear identification tags.

(g) Radiotelephone operators (RATELO) will use short whip antennas only. They will depress antennas to avoid breakage and to reduce the safety hazard. When directed, they will check communications with the tactical operations center, ensuring that the radio remains "on" during flight. They will have a minimum of two extra batteries for each radio.

(h) An accurate list for each aircraft load by name, grade, and unit will be furnished to the battalion SI, through the unit officer in charge of loading, for airloading table data.

(11) This sequence should be followed during the loading phase:

(a) Secure PZ.

(b) Approach aircraft only after it has landed.

(c) Do not load until station time.

(d) Load at double time.

(e) The aircraft troop commander establishes and maintains communication with parent headquarters, using his radio and unit net, upon landing and while in flight for changes in mission, LZs, or last-minute intelligence dissemination. If a unit radio is not available, he should receive this information from the crew through the helicopter radio.

(12) When loading personnel or cargo into a helicopter, the aircraft troop commander ensures that the following is accomplished:

(a) All safety measures prescribed for movement in and about the helicopter are observed.

(b) All personnel approach the helicopter in the prescribed manner.

(c) Personnel will not go near the tail rotor.

(d) All personnel and equipment will stay below the are of the main rotor. Personnel should be especially watchful when loading on the slope of a hill; approach and depart helicopter on downslope side. However, entering and exiting the helicopter should be made on the uphill side so the pilot can better control the helicopter. Personnel with backpacked radios will hold antennas down during any movement around aircraft.

(13) Briefing on emergencies will be conducted by an aviation representative prior to loading, as appropriate.

(14) After all equipment and personnel have been loaded, the air craft troop commander determines the following:

(a) Equipment and cargo are in the proper places.

(b) Cargo or equipment is properly secured.

(c) Each soldier is seated and his safety belt fastened.

(d) Weapons are placed between legs; muzzle up in utility aircraft muzzle down in cargo helicopters.

(15) When the aircraft troop commander has checked to ensure that all cargo and personnel are secured, he will notify the aircraft commander.

(16) During flight, the pilot commands the aircraft. The aircraft troop commander ensures that the following is accomplished:

(a) Cargo lashings (if applicable) are checked frequently to determine that cargo is properly secured.

(b) Troops keep safety belts secured and do not smoke unless authorized. (c) Troops stay seated and do not move around in the cabin without authorization.

(17) In the event more than one lift is required, the PZCO will remain until the last lift to ensure control and continuous communications.

(18) General aircraft load planning.

(a) All units will develop general load plans to facilitate movement on short notice.

(b) Necessary equipment, for aircraft loading and movement (nets, slings, and clevises), will be kept on hand.

(c) Personnel will be organized and trained in loading equipment to include slingloads.

(d) Battalion S3 Air will prepare airloading tables for movement by US Air Force aircraft. Companies will maintain airloading tables for air assault operations (Appendix B).

(e) Vehicles and major equipment will be prepared at all times to facilitate airlift operations. Chalk numbers will be predetermined for vehicles and trailers. Vehicles and major equipment to be transported into objective area will be reported with strength figures.

(f) A-22 containers: maximum height (max ht) 40 meters; max weight (wt) 1,200 pounds (lb) (for all helicopter operations). This will allow one container lift by UH-60s. If CH-47s are available, two or more A-22s may be rigged together.

e. Subsequent Operations.

(1) Withdrawal by air. Withdrawal from an area of operation requires thorough planning, close coordination, and controlled execution. The following are considered important for any withdrawal by air:

(a) Primary and alternate PZs and flight routes must be planned.

(b) Defensive concentrations must be planned around the PZ. The security force will protect the loading force and return fire if engaged. When the last elements are ready to load, the security force will call in required fires to cover withdrawal and use their own fire to cover their loading.

(c) Maintain all-round security until the first helicopter is on the ground (never assemble too early).

(d) Plan the loads so that a force capable of defending itself constitutes the last lift (never leave less than a platoon-size force). A platoon leader or sergeant, or squad leader, with radio, will be the last man out of a PZ. He will report to his commander that the PZ is clear of all personnel and equipment and immediately notify the pilot of the helicopter he boards. Plan for at least two extra helicopters to go into the PZ to lift out the last unit, when possible.

(e) The attack helicopter unit will be in direct communication with the ground force commander.

(2) Displacement of command post.

(a) Quartering party.

- Composed of Sl or headquarters commandant, communications officer or representative and communications personnel, security element, and other necessary personnel.
- Selects location.

(b) Command and control helicopter will be utilized as main CP during movement.

(c) Quartering party duties upon landing include:

- Laying out new CP.
- Notifying old CP when new CP is ready for operation.
- Ensuring timely and orderly arrival and positioning of other CP elements.
- Opening new CP. Officer in charge notifies commander or S3 when old CP has closed and when staff is operational in the new location.
- Controlling responsibilities. Command and control helicopter will be used as tactical CP to control and direct subordinate elements during air movement. The old CP is responsible for the dissemination of information and reports to higher and adjacent headquarters until that function is formally passed to the new CP.

(3) Security of aircraft in unit areas.

(a) Passive measures.

- Laagers (occupancy, 1-36 hours):
  - Select proper terrain for laagers where access by enemy ground forces is difficult (for example, laagers surrounded by water or swamps).
  - Site aircraft to blend with terrain and vegetation (locate parking areas in shadows, near trees).
  - Park aircraft in laagers so that attack helicopters can provide security along avenues of approach. Lift-off of aircraft, if attacked by enemy, is the responsibility of the aviation commander.
  - Utilize troops in or near the laagers to provide perimeter security. Aviation unit will augment security.
- Semipermanent facilities (occupancy, one to several weeks).
  - Use camouflage nets and natural materials to provide concealment.

- Provide perimeter troop security around airfields and heliports.
- Construct individual and helicopter bunkers and continue progressive improvement as time permits.

(b) Security of supporting aviation is the responsibility of the unit commander in whose area they are laagered or as designated by this headquarters.

5. LOGISTICS.

a. Supply.

(1) Accompanying supplies - all classes. Prescribed supplies will be established by this headquarters for each air assault operation.

(a) Class I. Each soldier will carry three combat ration meals to be eaten on order.

(b) Class II and IV. Units take on one-day supply of required combat essential expendables.

(c) Class III.

- Vehicle fuel tank, three-fourths full; gas cans, filled to the weld.
- Units take one-day supply of oil and lubricants on vehicles.

(d) Class V.

- Units maintain basic load at all times.
- Available supply rates and priority of delivery specified in OPORD.

(e) Class IX. Units take combat essential prescribed load list.

(f) Water. Soldiers carry two full canteens and one bottle of water purification tablets.

(2) All classes of supply delivered using unit distribution.

(3) Routine, planned supplies will be prepackaged to maximum extent possible by S4.

(4) Emergency resupply containing ammunition, water, rations, and medical supplies will be prepackaged by the S4 and will be ready for delivery as required.

b. Salvage.

(1) Expedite recovery of aerial delivery containers, parachutes, cargo nets, and pallets; commanders guard against damage, destruction, or loss.

(2) Units in objective area establish salvage collecting points when appropriate and practical.

(3) Salvage will be reported to this headquarters for disposition instructions.

c. Captured Materiel. Captured materiel may be used on approval of this headquarters. Evacuation of captured material is accomplished, as the situation allows, through S4 channels.

d. Health Service Support.

(1) Medevac of patients, until linkup or withdrawal, will be by air.

(2) Aid station location will normally be in battalion rear.

(3) Units report capture of medical supplies to battalion medical platoonpersonnel.

(4) Requests for medical evacuation within the AATF operations area will be to the medical organization on the medical evacuation frequency or the admin-log net. Requests for medical evacuation to division medical facilities will be submitted to the AATF S1.

(5) PW casualties needing medical treatment will be evacuated through medical channels.

(6) Hospital locations will be announced for each operation.

e. Transportation and Troop Movement.

(1) Vehicular.

(a) Allocation of accompanying organic transport will be made by this headquarters.

(b) Captured vehicles will be used to the maximum to meet transportation requirements.

(2) Aircraft. Allocation of supporting aircraft will be made by this headquarters.

6. VISUAL AND SOUND SIGNALS.

a. Use as required and prescribed by CEOI and unit SOP and as modified by battalion OPORD.

b. Subordinate units employ only those pyrotechnics specifically authorized by OPORD or CEOI.

7. ELECTRONIC WARFARE.

a. Radio stations will not attempt to enter, jam, or otherwise interfere with unknown radio nets without prior approval of this headquarters.

b. Report (by a secure means) jamming or attempts to enter nets by unknown stations to the communications officer without delay. Give time, radio frequency, type of jamming, signal strength, readability, and identity (if obtainable) of interfering station.

COLEMAN
LTC

OFFICIAL:

/s/BAXTER
BAXTER
S3

Distribution: A

Enclosures:

1 - Air Assault Formations and Battle Drill.
2 - Air Assault Task Force Commander's Checklist
3 - Air Mission Commander's, or Air Liaison Officer's, Checklist.

______________________(CLASSIFICATION)

ENCLOSURE 1 TO SOP FOR AATF OPERATIONS
AIR ASSAULT FORMATIONS AND BATTLE DRILL

1. GENERAL

This enclosure prescribes the PZ and LZ aircraft formations used when conducting air assault operations. It further prescribes the immediate action to be followed when personnel dismount at the LZ.

2. APPLICATION

This enclosure applies to all organic and supporting units under the control of this battalion. All subordinate SOPs conform, and all personnel will be thoroughly familiar with these procedures.

3. FORMATIONS

Aircraft supporting this battalion may use any of the following PZ, LZ configurations, as prescribed by the AATF commander working in conjunction with the AMC.

HEAVY LEFT OR RIGHT
Requires a relatively long, wide landing area; presents difficulty in pre-positioning loads; restricts suppressive fire by inboard gunners; provides firepower to front and flank.

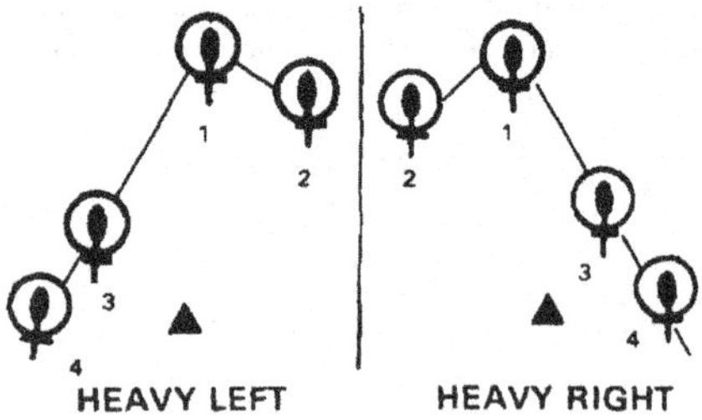

DIAMOND

Allows rapid deployment for all-round security; requires relatively small landing area; presents some difficulty in pre-positioning loads; restricts suppressive fire of inboard gunners.

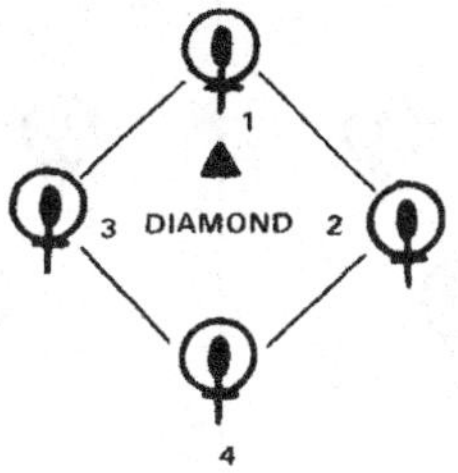

VEE

Requires a relatively small landing area; allows rapid deployment of forces to the front; restricts suppressive fire of inboard gunners; presents some difficulty in prepositioning loads.

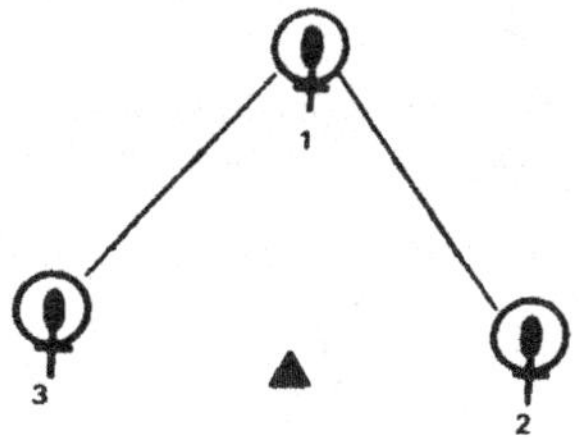

ECHELON LEFT OR RIGHT

Requires a relatively long, wide landing area; presents some difficulty in prepositioning loads; allows rapid deployment of forces to the flank; allows unrestricted suppressive fire by gunners.

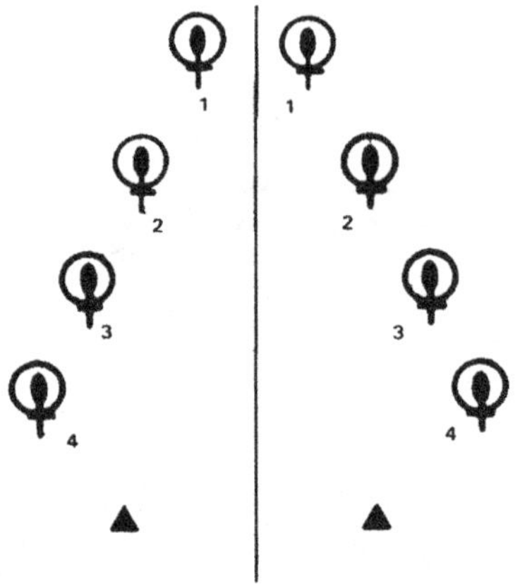

TRAIL
Requires a relatively small landing area; allows rapid deployment of forces to the flank; simplifies pre-positioning loads; allows unrestricted suppressive fire by gunners.

STAGGERED TRAIL LEFT OR RIGHT
Requires a relatively long, wide landing area; simplifies pre-positioning loads; allows rapid deployment for all-round security; gunners' suppressive fire restricted somewhat.

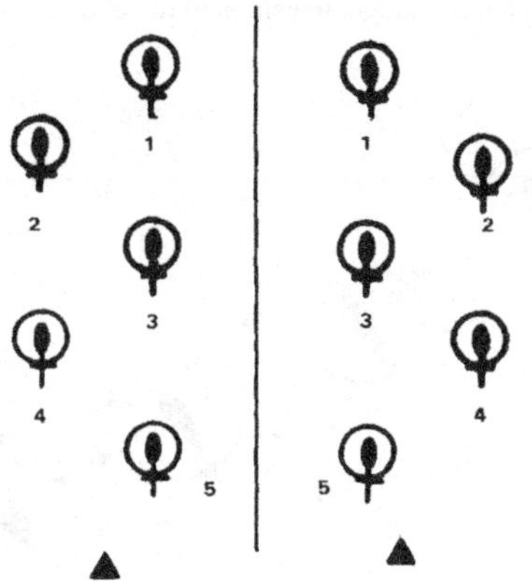

BATTLE DRILL

The first step in successful execution of air assault battle drill is to ensure that the aircraft is loaded so that dismounting soldiers react promptly and contribute to mission accomplishment. Regardless of the formation used, individual aircraft are always loaded as follows:

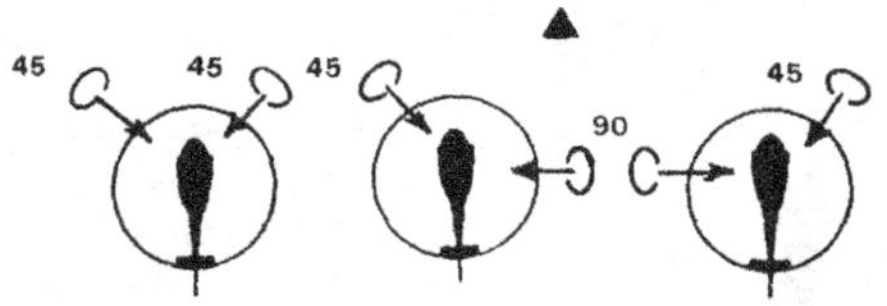

Dismounting in the LZ is the reverse of loading.

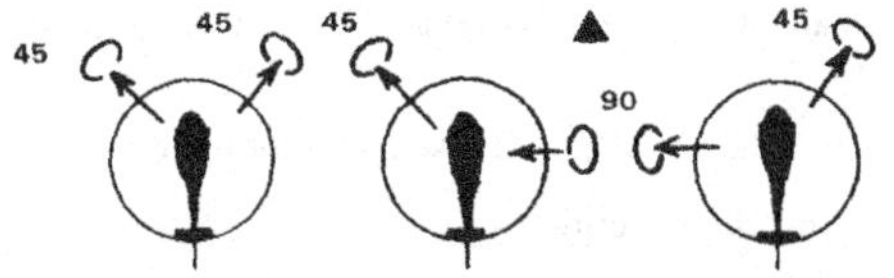

This method of loading and unloading is altered when aircraft are landing on a moderate slope.

In so doing, all-round security is attained in both the PZ and LZ, and the need for complicated procedures is reduced by keeping the positions for men and equipment the same. On the LZ, this facilitates use of the bounding overmatch method of movement. Bounding overmatch may be initiated in any direction.

From the initial bounding overwatch formation, transition to traveling overmatch is simple.

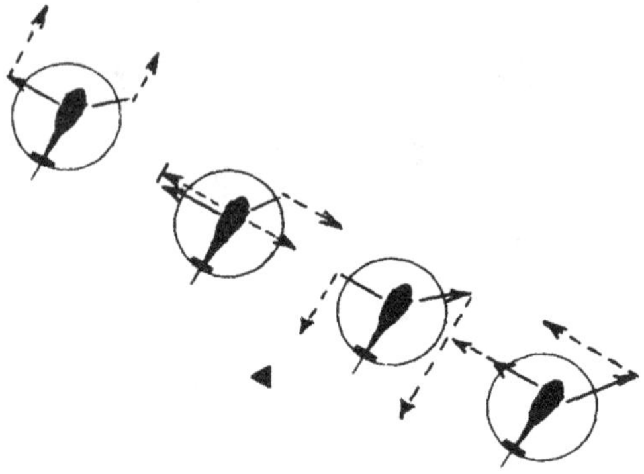

Broken lines indicate movement by bounding overmatch after departure of aircraft.

Dismounting into bounding overmatch.

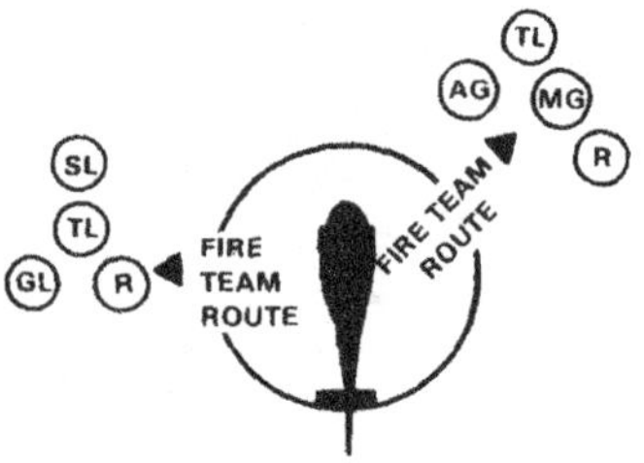

LEGEND

SL — Squad Leader
TL — Team Leader
GL — Grenade Launcher
R — Rifleman
MG — Machine Gunner
AG — Assistant Machine Gunner

Squad organizes to fit eight-man ACL. (If the CH-60 Black Hawk is used, the remaining squad members can be included in formation.)

If enemy contact is not expected, the squad can be assembled for traveling overmatch.

Dismounting and changing to traveling overmatch.

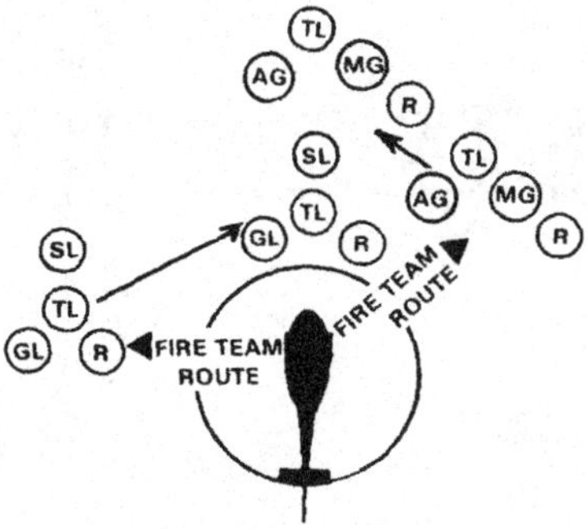

Movement to a line or assault formation can be executed. (Additional squad members can be carried on Black Hawk.)

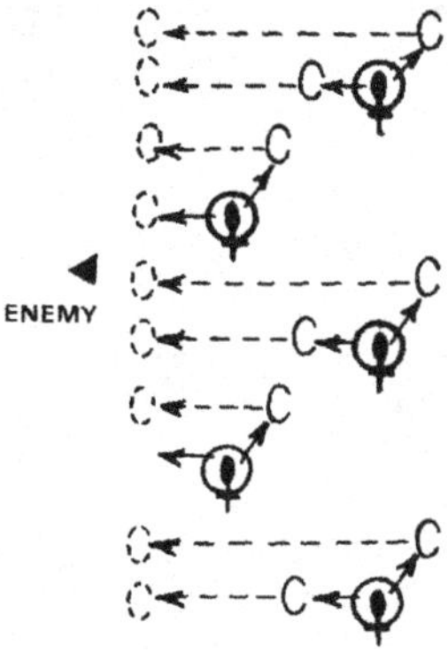

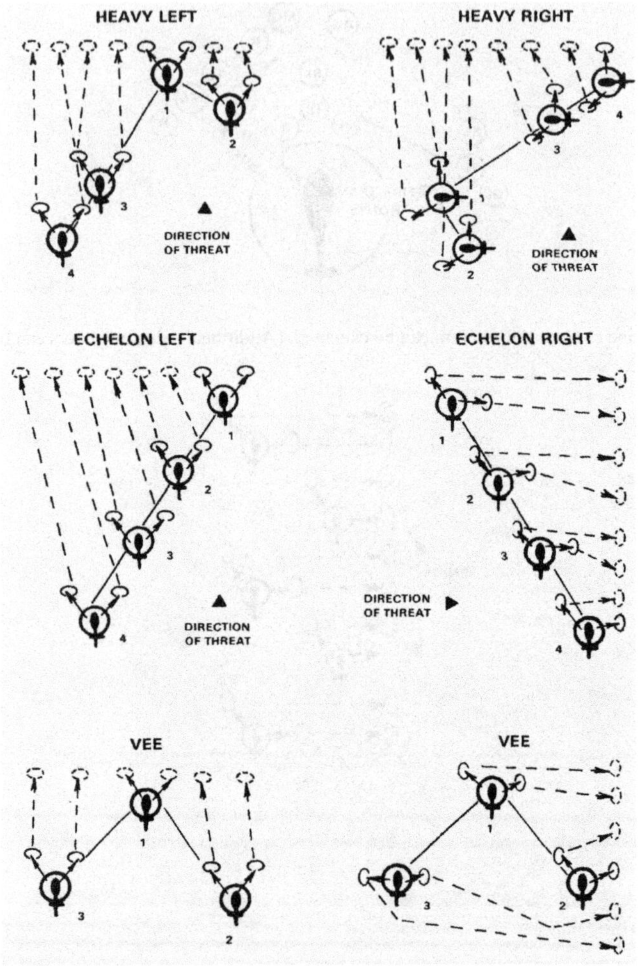

Broken lines indicate movement after aircraft departure.

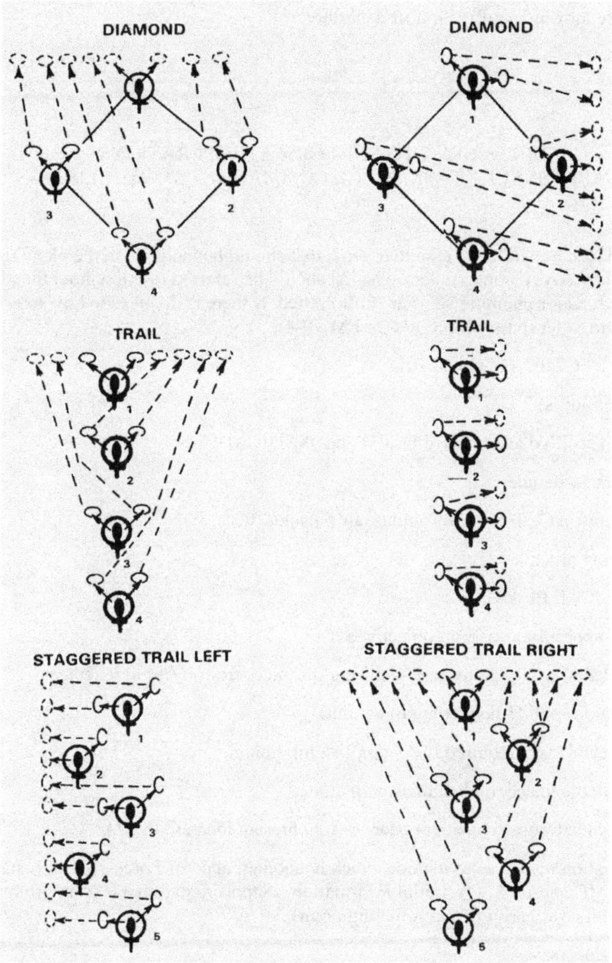

Broken lines indicate movement after aircraft departure.

Broken lines indicate movement after aircraft departure.

ENCLOSURE 2 TO SOP FOR AATF OPERATIONS
AIR ASSAULT TASK FORCE COMMANDER'S CHECKLIST

1. GENERAL

This list is designed to summarize the essential items that should be included in the planning phase of an air assault operation by AATF commanders. The list should be referred to throughout the planning process to ensure that major planning steps are not omitted. If there is doubt as to how to accomplish a particular task or item, refer to the unit SOPs (or FM 90-4).

2. ACTION UPON RECEIPT OF ORDERS

a. Analyze mission(s).

b. Determine specified and implied task(s) and objectives).

c. Develop time schedule.

d. Obtain aircraft ACL from AMC and/or air liaison officer.

e. Issue warning order.

3. GROUND TACTICAL PLAN

a. Choose, as appropriate, assault objectives.

b. Designate LZ(s) available for use. Consider distances from LZ(s) to objective.

c. Establish D-day and H-hour (time of assault).

d. Identify special tasks required to accomplish mission.

e. Means available to accomplish mission include:

(1) Organic troops (consider distance from present location to PZ),

(2) Aviation resources to include attack helicopter, and Air Force support (establish liaison with AMC and/or ALO) (initial information, support requirements from ground unit to include forward arming and refueling point).

(3) Engineers.

(4) Signal to include aerial radio relay.

(5) Medical.

(6) Fire support.

(a) Close tactical air support.

(b) Field artillery within range.

(c) Other indirect fire weapons (mortar and naval gunfire).

(d) Preparation fires for LZs (signals for lifting and/or shifting).

(e) Flight corridors.

(f) Air defense suppression.

(7) Control measures needed.

(8) Subsequent operations (for example, defense linkup, withdrawal) that may be conducted.

NOTE: Announceconcepttostailandsubordinatounitsassoonaspossibletofaciliteteplanning.

4. INTELLIGENCE INFORMATION AND REQUIREMENTS

a. Enemy locations to include air defense positions.

b. Commander's aerial recon of objective area (if practical).

c. Aerial reconnaissance (side-looking airborne radar [SLARI aerial photos).

d. Sensor reports.

e. Terrain study.

f. Weather forecast.

g. Latest intelligence summary (INTSUM).

h. EPW handling procedures.

i. Civilian control procedures.

5. LANDING

a. Selection of primary and alternate LZs (capacity).

b. LZ identification procedures for landing sites include:

(1) Colored smoke.

(2) Panels.

(3) Flares.

(4) Lights.

c. Use of pathfinders.

d. Landing formations).

e. Approach and departure directions.

f. LZ preparation fires to support landing plan and ground tactical plan include:

(1) Use of TACAIR (close air support, air defense suppression, and air cover).

(2) Use of indirect fire weapons.

(3) Use of attack helicopters and units.

(4) Use of EW.

g. Other fire support considerations include:

(1) Shifting of fires.

(2) Lifting of fires.

(3) Suppression of enemy air defenses.

6. AIR MOVEMENT

a. Flight routes (primary-alternate-return) require the following data:

(1) RPs; direction and distance to LZs.

(2) SP; air control points, CCP, and RPs.

(3) Phase lines (if used).

(4) Estimate time en route.

(5) Maneuver areas for attack helicopter.

(6) Laagers, to include location, mission, and security.

(7) Friendly air defense considerations.

(8) Enemy air defense intelligence.

b. Air movement table to implement air movement includes:

(1) Units to be lifted.

(2) Number and type of lift helicopters allocated to each unit.

(3) Aviation units that will support unit.

(4) Lift-off times.

(5) Routes.

(6) Unit LZs.

(7) H-hour (landing time of initial serial).

c. Alternate communications plan includes:

(1) FM.

(2) UHF.

(3) VHF.

(4) Visual/audio signals.

(5) Aerial radio relay.

7. LOADING

a. PZ assignment by unit (primary-alternate) (bump and/or straggler contingency plan).

b. Holding areas.

c. Routes from assembly areas to holding area to PZ(s).

d. Attack helicopter utilization (overmatch and security) includes:

(1) En route to PZ.

(2) While lift aircraft are in PZ.

(3) En route to LZ.

(4) Recon of LZ; marking of LZ.

8. SUPPORT PIANS FOR CONDUCT OF AIR ASSAULT OPERATIONS

a. Alternate plans and procedures due to weather (H-hour increment to delay operation).

b. Downed helicopter procedures to include:

(1) Crew and passenger duties.

(2) Aircraft disposition instruction.

c. Rally points.

d. Escape and evasion instructions.

e. Laager sites.

f. Rules of engagement.

g. Deception plans that will be used.

h. Spare aircraft available.

i. Reconnaissance (air-ground) that will be conducted.

j. Straggler control procedures.

k. Reporting (en route, lift-off, touchdown, intelligence, and contact).

l. Aircraft disposition after assault.

m. Health service support and evacuation procedures.

9. ACTIONS THAT MUST BE COMPLETED

a. Warning orders.

b. Liaison officer (receive and dispatch).

c. Attachments and detachments.

d. Issue commander's concept (time and place).

e. Briefings (time and place).

f. Preparation of OPORD.

g. Issue OPORD (time and place).

10. LOGISTICS REQUIREMENTS

a. Class V resupply.

b. Feeding plan.

c. Water.

d. Medevac (call sign, frequency, location, and procedures).

e. Refueling (location of FARP, ammunition available).

11. DEBRIEFING

a. Lessons learned:

(1) Ground units.

(2) Aviation units.

b. Actions taken for correction.

ENCLOSURE 3 TO SOP FOR AATF OPERATIONS

AIR MISSION COMMANDER'S
OR AIR LIAISON OFFICER'S CHECKLIST

1. GENERAL

This list is designed to summarize the essential items included in the planning phase of an air assault operation by the AMC. The list is referred to throughout the planning process to ensure that major planning steps are not omitted. If there is doubt as to how to accomplish a particular task or item, refer to the unit SOPs (or FM 90-4).

2. ACTION PRIOR TO DEPARTURE FOR SUPPORT UNIT

Meet attack helicopter and pathfinder representatives at prearranged site. Obtain briefing from operation's officer to include:

a. Support unit.

(1) Mission.

(2) Location.

(3) Contact officer.

(4) FM frequency.

(5) Call sign.

b. Mission.

(1) Requirements for aerial reconnaissance.

(a) Utility helicopters.

(b) Attack helicopters.

(c) Observation helicopters.

(2) Special mission requests.

(3) Number of aircraft, by type, that are available for the operation (status of assets).

(4) Utility, observation, cargo, or scout helicopters.

(5) Attack helicopters.

c. ACL for each type of aircraft.

(1) Number of troops ______; pounds of cargo ______ .

(2) Number of pathfinders available and time available.

(3) Pathfinder equipment available.

(4) Specific problem areas or requirements that may affect support of ground unit (FARP location and time of operation). (Estimated refueling time, and refuel-rearm plan.)

(5) Obtain necessary equipment that will be required at or by supported unit.

(a) Aircraft or vehicle.

(b) Maps, overlays, photographs.

(c) Radios, CEOI for exchange.

(d) Personal gear

(e) Additional headsets for reconnaissance, if required.

(6) Check with AATF commander for special instructions.

NOTE: Confirm if supported unit is prepared to receive LO.

3. ACTIONS EN ROUTE

a. Establish and maintain communications.

b. Obtain status of fires and permission to enter area of operations.

4. ACTIONS AT SUPPORTED UNIT LOCATION

a. Report to supported commander, S3, or LO.

b. Brief supported unit on number and type of aircraft available, ACL, and other essential information.

c. Obtain initial briefing on the following:

(1) Enemy situation.

(2) Friendly situation.

(3) Ground tactical plan (make map overlays).

(4) Supported participating aviation units. Coordinate and integrate plans as necessary.

d. Assist supported unit in planning the following:

(1) Movement to PZ for ground and aviation unit and control facilities.

(2) Loading.

(a) Location and selection of PZ.

(b) Special PZ marking procedures.

(c) Aircraft marking procedures.

(d) Landing formation and direction.

(e) Loads:

- Troops.
- Cargo.

(f) Communication control procedures.

(g) PZ control (obtain call sign and frequency).

(h) Manifesting (completion of airloading table)

(i) Priorities of bump by aircraft.

(j) PZ and lift-off times.

e. Air movement.

(1) Flight route. Provide guidance and give technical approval oon selection of the following

(a) SP.

(b) ACPS.

(c) CCPS.

(d) RP.

(2) Alternate and return flight routes.

(3) Formation: select en route formation that gives the most control and is least vulnerable to enemy interference; provide guidance for selection of PZ and LZ formations.

(4) Altitude and speed.

(5) Overwatch and security plan for attack and scout helicopters.

(6) Fire support plan en route.

(7) Air movement table (assist in completing).

(8) Pathfinder support (finalize)

f. Landing.

(1) Touchdown time(s) (in terms of H-hour).

(2) LZ designations and locations.

(3) Size and description.

(4) LZ marking and procedures.

(5) Landing directions.

(6) Landing formation.

(7) Traffic pattern for subsequent lifts.

(8) Communications, control procedures, and use of pathfinders.

g. LZ preparatory and suppressive fires.

(1) CAS (start time, duration, target and type of ordnance, and attack direction).

(2) Indirect fires (start time, duration, target and type of fuze, special instructions).

(3) Plan for attack helicopter unit's scheme of maneuver and plan for overmatch and security (start time, duration, special instruction, attack direction).

(4) Firing of lift helicopter weapons (provide suppressive fires upon landing).

(5) Fire plan of debarking troops.

(6) Call signs and/or frequency signals for lifting and/or shifting support fires.

h. Refueling requirements.

(1) Location of FARP(s).

(2) Time required.

i. Aircraft maintenance.

(1) Downed aircraft procedures.

(2) Spare aircraft procedures.

5. ACTIONS PRIOR TO DEPARTURE FROM SUPPORTED UNIT

a. Obtain copies of OPORD with overlays and annexes.

b. Confirm all times.

c. Last-minute weather check.

(1) Mission procedures (delay increments).

(2) Alert procedures.

d. Debriefing the commander.

6. ACTIONS UPON RETURN TO AVIATION UNIT

a. Inform unit commander.

b. Brief personnel, as appropriate, on all above information.

c. Maintain close liaison with support unit.

EXECUTE MISSION AS PLANNED

MISSION DEBRIEFING FOR AVIATION UNIT

AFTER-ACTION REPORT

APPENDIX D

Helicopter Characteristics

D-1. General.

a. This appendix discusses the characteristics of US Army aircraft (Chart D-1)

b. To efficiently load an AATF aboard helicopters, commanders and staffs must know the exact composition of the AATF, the essential characteristics of the types of helicopters to be used for the operation, and the methods of computing aircraft requirements.

c. Maximum ACIis are affected by altitude and temperature and will differ widely according to topography and climatic conditions common to specific zones or areas of military operations. ACLs will further vary based on the location of, approaches to, and exits from LZs; pilot proficiency; aviation unit SOP; type of engine in the aircraft; and age of both aircraft and aircraft engine. Therefore, two identical aircraft, of the same model and type, may not be able to pick up and carry identical loads.

D-2. References.

TM 55-450-15 provides detailed characteristics of Army aircraft, technical data, and guidance for computing aircraft requirements, and examples of detailed airloading and air movement forms.

D-3. Aircraft availability.

a. Aircraft availability is an overriding consideration in air assault operations. It is directly influenced by the adequacy and efficiency of maintenance and supply activities, and aircraft utilization and scheduling procedures, as well as by the distance of support units from the operating units.

b. Both the support and supported commanders should be aware that everyday use, over an extended period, of all available aircraft will result in a reduced mission availability rate for future operations. During periods of sustained operations, fixed-wing aircraft normally can maintain a greater percentage of aircraft availability for longer periods than helicopters. In the course of sustained operations, aircraft maintenance must be carefully considered and programmed so that heavy flying requirements will not cause a continual decrease in aircraft availability.

c. Supported unit commanders, staffs, and logistical planners must conserve the use of available aircraft by:

(1) Establishing acceptable availability rates prior to operational commitment.

(2) Establishing forward refueling and/or rearming areas to eliminate flying hours expended for those purposes.

(3) Utilizing surface means of transportation for logistical support whenever possible.

(4) Timely and coordinated logistical planning to ensure full utilization of all aircraft sorties and to avoid duplication of effort.

D-4. Capabilities and limitations of army aircraft.

a. **Helicopters.**

(1) **Capabilities.**

(a) Under normal conditions, helicopters can ascend and descend at relatively steep angles, a capability which enables them to operate from confined and unimproved areas.

(b) Troops and their combat equipment can be unloaded from a helicopter hovering a short distance above the ground with troop ladders and rappelling means, or if they can hover low enough, the troops may jump to the ground. The troop ladder can also be used to load personnel when the helicopter cannot land.

(c) Cargo can be transported as an external load and delivered to areas inaccessible to other types of aircraft or to ground transportation.

(d) Because of a wide speed range and high maneuverability at slow speeds, helicopters can fly safely and efficiently at a low altitude, using terrain and trees for cover and concealment.

(e) Their ability to fly at high or low altitudes and to decelerate rapidly, combined with their capacity for slow forward speed and nearly vertical landing, enables helicopters to operate under marginal weather conditions.

(f) Helicopters can land on the objective area in a tactical formation, LZ(s) permitting.

(g) Night and/or limited visibility landings and lift-offs can be made with a minimum of light.

(h) Helicopters flying at low levels are capable of achieving surprise, deceiving the enemy at the LZ(s), and employing shock effect through the use of suppressive fires.

(i) Engine and rotor noise may deceive the enemy as to the direction of approach and intended flight path.

(2) **Limitations.**

(a) The high fuel consumption rate of helicopters imposes limitations on range and ACL. Helicopters may reduce fuel load to permit an increased ACL. However, reducing the fuel load reduces the range and flexibility factors, which must be considered in planning.

(b) Weight and balance affect flight control. Loads must be properly distributed to keep the center of gravity within allowable limits.

(c) Hail, sleet, icing, heavy rains, and gusty winds (30 knots or more) will limit or preclude use of helicopters.

(d) Engine and rotor noise may compromise secrecy.

(e) Aviator fatigue requires greater consideration in the operation of rotary-wing aircraft than in the operation of fixed-wing aircraft.

(f) The load-carrying capability of helicopters decreases with increases of altitude, humidity, and temperature. This limitation may be compensated for through reduction of fuel load.

(g) Crosswind velocities above 15 knots for utility and 10 knots for cargo helicopters, and downwind velocities above 5 knots for either type of helicopter, will affect the selection of the direction of landing and lift-off.

b. **Fixed-wing aircraft.**

(1) **Capabilities.**

(a) Army fixed-wing aircraft can operate from relatively short LZ(s) if the terrain is fairly smooth and level.

(b) Fixed-wing aircraft have a greater range than rotary-wing aircraft and require less maintenance.

(c) On some fixed-wing aircraft, cargo can be transported as an external load suspended from bomb shackles on the wings and can be dropped with a high degree of accuracy from low altitudes.

(d) Landings and takeoffs at night or during limited visibility can be made with a minimum of ambient light.

(2) **Limitations.**

(a) Fixed-wing aircraft may require improved landing strips.

(b) Hail, sleet, icing, heavy rains, and gusty winds (30 knots or more) will limit or preclude use of these aircraft.

(e) A wind velocity of 8 to 10 knots affects the selection of the direction of landing and takeoff.

c. **Special considerations.** The capabilities and limitations of rotary-wing and fixed-wing aircraft are variable. Commanders of supporting aviation units provide specific data for each type of aircraft and operation.

D-5. Methods of determining army aircraft requirements.

a. **Weight method** (Chart D-1).

(1) The weight method is used when the total weight to be transported is the determining factor. However, this method is not accurate enough to compute requirements for units that must transport major items of equipment and also maintain tactical integrity. Aircraft requirements are determined by dividing the ACL (payload) of each aircraft into the total

weight of the force to be airlifted. Whenever the weight method is used, care must be taken so that any one load does not exceed the ACL of the aircraft being used and that any one piece of equipment is within the size and weight limit of the aircraft. This method of aircraft estimation is not particularly accurate and should not be used below division level.

(2) Example of UH-1 using the weight method.

(a) Total weight to be transported 60,970 pounds.

(b) Allowable cargo load 1,700 pounds.

(c) 60,970 divided by 1,700 equals 35.8 or 36 UH-1s.

b. **Type-load method.**

(1) The type-load method is the most efficient method to be used in the conduct of air assault operations and in operational planning. Army aviation units are frequently required to support numerous major units operating over expansive tactical zones. Standardization of type loads within the theater of operation ensures responsive and effective airmobility with a minimum of time required for planning. The use of type loads does not limit the flexibility of a ground tactical unit to be airlifted. The type-load method can also be used at battalion and company levels to plan and conduct air assault and joint airborne operations.

(2) The use of type loads provides a drill-type SOP operation, thereby reducing the time required for planning and computation, and reducing the confusion and error common to air assault operations conducted with minimum advanced notification.

D-6. Sample load types for army rotary-wing aircraft.

a. **UH-1**. Maximum ACL for the UH-1 is 2,000 pounds. As fuel is reduced following the initial airlift, the troop load may be increased to eight or nine for subsequent lifts.

	Cargo	**Weight** (lbs)	**Total**
(1)	7 personnel	1,680	1,680
(2)	Bulk cargo	2,000	2,000
(3)	1 ea Mule (slingload)	900	
	Load on Mule	1,000	1,900
(4)	1 ea 1/4-ton trailer	565	
	Load on trailer (external load)	500	1,065

b. **UH-60.** Maximum ACL is 11 to 13 troops, depending on the aircraft seating configuration.

	Cargo	**Weight** (lbs)	**Total**
(1)	11 personnel	2,640	
	1 ea 1/4-ton truck with trailer	3,500	6,140
(2)	7 personnel	1,680	
	1 ea M102 howitzer	3,195	
	40 rds ammo (A-22)	2,400	7,725

c. **CH-47.** Type-load data is based on an aircraft maximum gross weight of 33,000 pounds on a standard day at mean sea level. As density altitude increases, or when the aircraft is required to operate at higher altitudes, the payload is reduced accordingly.

	Cargo	**Weight** (lbs)	**Total**
(1)	20 personnel	4,800	
	1 ea A-22 container (slingload)	3,000	7,800
(2)	8 personnel		
	Mules (loaded)	6,000	7,920
(3)	22 personnel	5,280	
	3 ea 81-mm mortars	282	
	150 rds ammo (slingload)	2,250	7,812
(4)	16 personnel	3,840	
	2 ea mortars 4.2-inch	1,200	
	100 rds ammo (slingload)	3,000	8,040
(5)	6 personnel	1,440	
	1 1/4-ton truck with 1/4-ton trailer	3,870	
	(slingload)	3,000	7,940
(6)	6 personnel	1,440	
	2 1/4-ton truck	5,200	
	1/4-ton trailer with load	1,000	7,640
(7)	3 personnel	720	
	1 M101A1 howitzer with sec equip	4,680	
	40 rds ammo	2,400	7,800
(8)	2 personnel	480	
	1 ea 3/4-ton truck	5,917	
	with cargo	1,500	7,897
(10)	3 personnel	720	
	1 1/4-ton truck with 1/4-ton trailer	3,870	
	1 Mule (loaded)	2,000	6,590
(11)	4 personnel	960	
	1 1/4-ton truck with MRC-95 radio	3,000	
	1 1/4-ton truck with 1/4-ton trailer	3,870	7,830
(12)	33 personnel	7,920	7,920
(13)	1 M102 howitzer	3,195	
	60 rds 105-mm ammo	3,600	
	Equip	430	7,225

d. **CH-54**. Type-load data are based on an aircraft maximum gross weight of 38,000 pounds on a standard day at mean sea level. As density altitude increases, or when the aircraft is required to operate at higher altitudes, the payload is reduced accordingly.

(1) Sample pod loads.

	Weight (lbs)
(a) Mixed cargo	10,000
(b) 1 3/4-ton truck with trailer	8,000
(c) 150 rounds 105-mm ammo (boxed)	17,000
(d) 7,200 C-rations, individual	12,550
(e) 67 troops at 240 lbs ea	16,080

(2) Sample 4-point slingloads.

	Weight (lbs)
(a) 2 1/2-ton truck	13,000
(b) Road grader (front see)	9,000
(c) Road grader (rear see)	14,000
(d) HD6 Bulldozer	16,000
(e) Personnel carrier M113	18,000

(3) Sample single-point slingloads.

	Weight (lbs)
(a) 4,500-gallon fuel bags	13,200
(b) CH-47 helicopter minus engines and blades	16,000
(e) CH-34 helicopter, complete	8,500
(d) OV-1 Mohawk	12,000
(e) 155-mm howitzer	14,000
(f) 100 rds 155-mm ammo	14,000

APPENDIX E

Pickup Zone and Landing Zone Operations

Section I

PLANNING PREPARATION

E-1. General.

This appendix serves as a small unit (company and below) leader's guide for the safe, efficient, and tactically sound conduct of operations in and around PZs and LZs.

E-2. Selection and marking.

Small unit leaders should be proficient in the selection and marking of PZs and/or LZs and in the control of aircraft.

a. **Tactical and technical considerations** (Chapter 3).

b. **Marking PZs and LZs.**

(1) **Day.** A ground guide will mark the PZ or LZ for the lead aircraft by holding an M16A1 rifle over his head, by displaying a folded VS-17 panel chest high, or by other identifiable means.

(2) **Night.** The code letter Y (inverted Y) is used to mark the landing point of the lead aircraft at night. Chemical light sticks or "beanbag" lights may be used to maintain light discipline (Figure E-1).

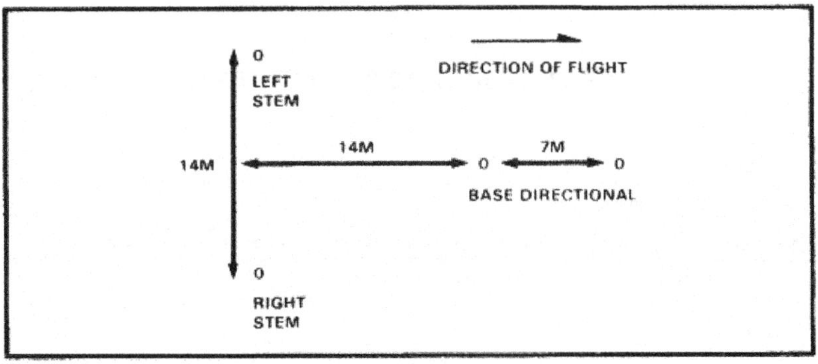

Figure E-1. Inveted Y.

When more than one aircraft will be landing in the same PZ or LZ, there will be an additional light for each aircraft. For observation, utility, and attack aircraft, each additional aircraft landing point will be marked with a single light emplaced at the exact point that each aircraft is to land. For cargo aircraft (CH-47, CH-53, CH-54), each additional landing point will be marked with two lights. The two lights will be placed 10 meters apart and will be aligned in the aircraft direction of flight.

(3) **Obstacles.** These include any obstruction to flight which might interfere with aircraft operation on the ground (trees, stumps, rocks) and cannot be reduced. During daylight, the aircrew is responsible for avoiding obstacles on the PZ or LZ. For night and limited visibility operations, all obstacles will be marked with red lights. The following criteria will be used in marking obstacles:

(a) If the obstacle is on the aircraft approach route, both the near and far sides of the obstacle will be marked.

(b) If the obstacle is on the aircraft departure route, the near side of the obstacle will be marked.

(c) If the obstacle protrudes into the PZ or LZ, but is not on the flight route of the aircraft, the near side of the obstacle **will** be marked.

(d) Large obstacles on the approach route will be marked by circling the obstacle with red lights.

c. **Control of aircraft.** Approaching aircraft are controlled by the use of arm-and hand signals to transmit terminal guidance for landing. The signalman is positioned to the right front of the aircraft where he can best be seen by the pilot. Signals at night are given by using lighted batons or flashlights in each hand. When using flashlights, care will be taken to avoid blinding the pilot. Batons and flashlights will remain lighted at all times when signaling. The speed of arm movement indicates the desired speed of aircraft compliance with the signal.

Section II

ASSEMBLY AND OBJECTIVE AREAS

E-3. General.

Prior to arrival of the aircraft, the PZ must be secured, PZ control party positioned, and the troops and equipment positioned in a unit assembly area.

a. **Occupation of unit assembly area.** Unit leaders should accomplish the following:

(1) Maintain all-round security of the assembly area.

(2) Maintain communications.

(3) Organize troops and equipment into chalks and loads in accordance with the unit air

movement plan.

(4) Conduct safety briefing and equipment check of troops.

(5) Establish priority of loading for each man and identify bump personnel.

(6) Brief on the location of the straggler control points.

b. **Movement to and occupation of chalk assembly area.** Linkup guides from the PZ control party will meet with designated units in the unit assembly area and coordinate movement of chalks to a release point. As chalks arrive at the release point, chalk guides will move each chalk to its assigned chalk assembly area. (To reduce the number of personnel required, the same guide may be used to move the unit from the unit assembly area to the chalk assembly area.) If part of a larger air assault, no more than three chalks should be located in the chalk assembly area at one time. Noise and light discipline will be maintained throughout the entire movement in order to maintain the security of the PZ. Additionally, no personnel should be allowed on the PZ unless loading aircraft, rigging vehicles for slingload, or directed by PZ control. While remaining in chalk order, each soldier is assigned a security (firing) position by the chalk leader and emplaced in the prone position, weapon at the ready, and facing out (away from PZ) to provide immediate close-in security.

(1) An example of a large, one-sided PZ is depicted in Figure E-2.

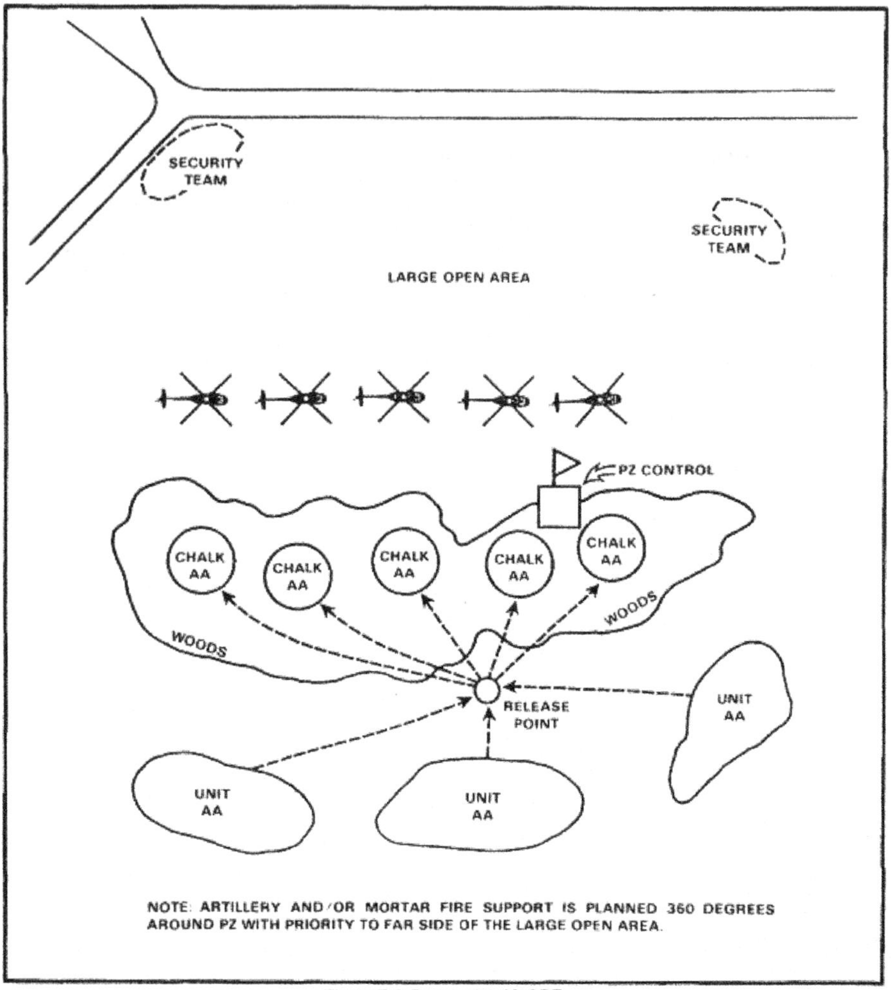

Figure E-2. Large, one-sided PZ.

(2) An example of a small two-sided PZ with unit and chalk assembly areas is depicted in Figure E-3.

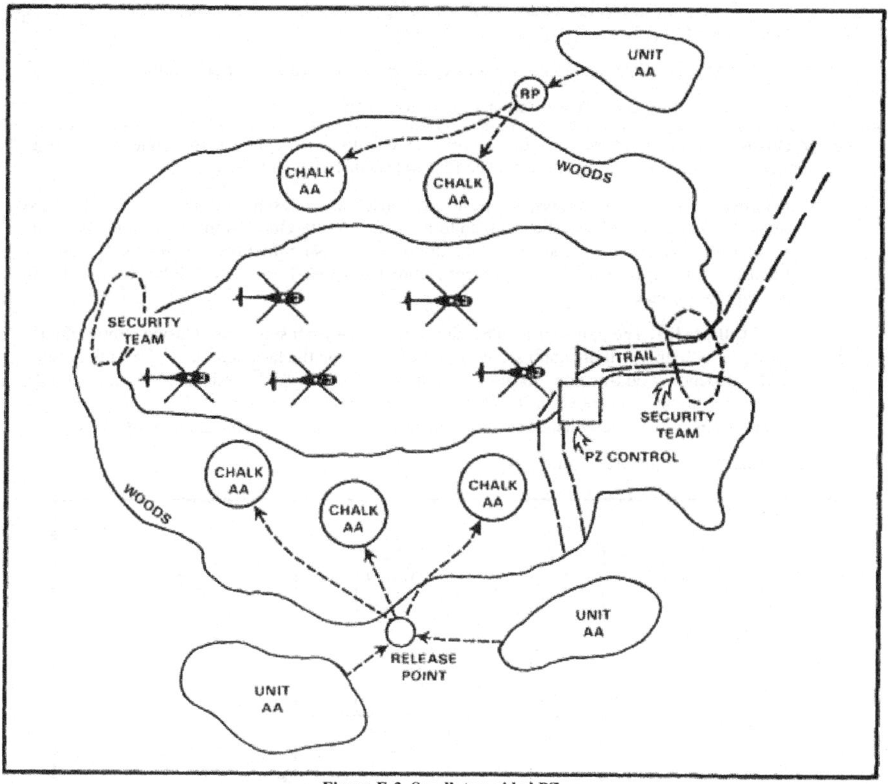

Figure E-3. Small, two-sided PZ.

(3) While in the chalk assembly area, units should adhere to the following principles for loading the aircraft:

(a) Maintain tactical integrity by keeping fire teams and squads intact.

(b) Maintain self-sufficiency by loading a weapon and its ammunition on the same aircraft (Dragon).

(c) Ensure key men, weapons, and equipment are cross-loaded among aircraft to prevent the loss of control, or all of a particular asset, if an aircraft is lost.

(d) Prior to loading, ensure all troop gear is tied down and checked; short antennas placed in radios, folded down, and secure.

(e) Squad and team leaders check the equipment of their men to ensure it is complete and operational.

(f) Radios on and communications check performed (unless directed otherwise).

(g) Specific aircraft seats are assigned to each man.

c. **PZ closure.** During platoon air assault operations the platoon sergeant is responsible for ensuring all personnel and equipment are loaded (clear the PZ) and security is maintained.

(1) **Single lift.** The platoon sergeant positions himself at the last aircraft and collects "bumped" men, if required. He will be the last man to load the aircraft. Once on the aircraft, the platoon sergeant (PSG) will notify the crew chief and/or AMC (using the troop commander's radio handset) that all personnel and equipment are loaded. Close-in security will be provided by the aircraft door gunners.

(2) **Multiple lift.** The duties of the PSG are the same as for a single lift. During a multiple lift, the security teams will maintain security of the PZ and be the last element to depart with the PSG. Depending on the initial locations) of the security teams, repositioning closer to the PZ may be necessary. Whenever possible, the aircraft will land as close to the security team positions as possible to enhance security and minimize the movement required by the teams.

d. **UH-60 loading secuence.** (Figure E-4).

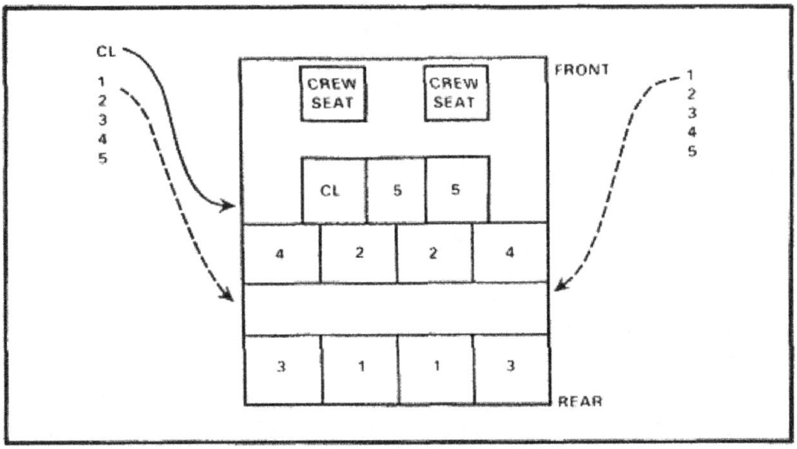

Figure E-4. UH-60 loading diagram, split chalk.

(1) Chalk leader (squad leader) initiates movement once the aircraft has landed.

(2) The farside and nearside groups move to the aircraft in file with the chalk leader (CL) always leading the nearside group.

(3) Chalk leader should:

(a) Ensure all personnel know which aircraft and which position to load.

(b) Ensure all personnel wear or carry rucksacks on the aircraft.

(c) Notify the crew chief when all chalk members are on board and are ready for lift-off.

(4) All personnel will buckle up as soon as they are seated in their assigned seats. The chalk leader will always sit in the left front seat unless a platoon leader or company commander is on the same aircraft.

(5) The chalk leader will hand the chalk card to the pilot and answer any questions the pilot may have, utilizing the aircraft intercommunication (troop commander's) handset.

e. **UH-60 loading sequence** (Figure E-5).

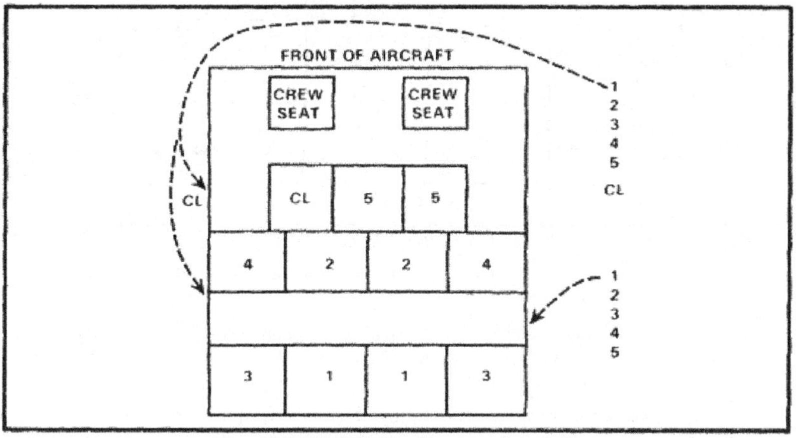

Figure E-5. UH-60 loading diagram, whole chalk.

(1) The chalk leader (squad leader) initiates movement once the aircraft has landed.

(2) The farside and nearside groups move to the aircraft in file with the No. 1 man leading the load to the appropriate side (Figure E-5).

NOTE: The farside group will always move around to the front of the aircraft.

(3) The chalk leader will stop at the nearside of the aircraft to ensure the nearside group loads properly; then he moves around the front of the aircraft to the farside and checks the other half of the chalk.

(4) All personnel will buckle up as soon as they are seated in the correct seat.

(5) The chalk leader will hand the chalk card to the pilot and answer any questions the pilot may

have, utilizing the aircraft's intercommunication (troop commander's) handset.

E-4. Landing zone operations.

Just as there is a priority of work for defensive operations, there is a priority of actions upon landing in an LZ.

a. **Unloading.** Unloading of the aircraft does not begin until directed by the crew chief or pilot (Figure E-6).

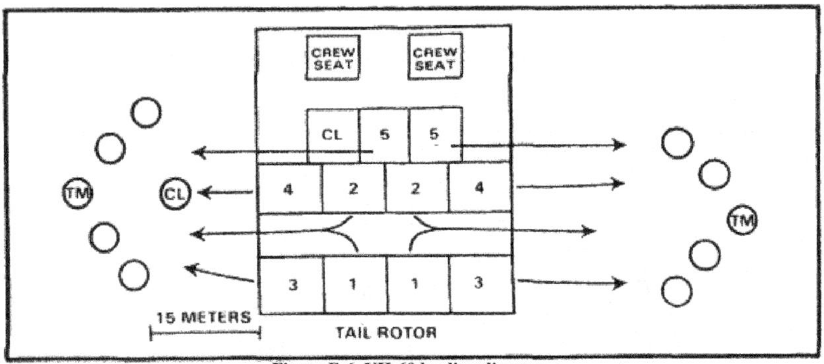

Figure E-6. UH-60 loading diagram.

(1)Once the aircraft has landed, personnel will unbuckle seatbelts and exit aircraft as fast as possible with all equipment.

(2) Prior to leaving the aircraft, the chalk leader will obtain the landing direction from the pilot if not determined during the approach into the LZ. This will aid in orientation to the LZ, particularly at night.

(3) Individuals will move 15 to 20 meters out from the side of the aircraft and assume the prone position facing away from the aircraft, weapons at the ready, until the aircraft has departed the LZ.

b. **Immediate action on hot LZ.** If the decision is made to use a hot LZ, or contact is made upon landing, troops quickly dismount and move 15 to 20 meters away from the aircraft and immediately return the enemy fire to allow the aircraft to depart.

(1) If the contact is similar to a far ambush, troops will fire and maneuver off the LZ to the closest side offering cover and concealment.

(2) If troops are engaged from nearby enemy positions, they treat it as a near ambush by immediately returning fire. Soldiers who consider themselves in the kill zone may assault the enemy positions or attempt to get out of the kill zone. Soldiers not in the kill zone will provide supporting fire to support the movement of troops in the kill zone.

(3) The squad or platoon leader will call for fire support if it is available.

(4) Once disengaged from the enemy force, the squad or platoon leader will move the unit to a covered and concealed position, account for personnel and equipment, and assess the situation as to whether or not the unit can continue the mission.

c. **Chalk assembly on cold LZ.** Upon unloading from the aircraft, the chalk leader (squad leader) will move the chalk to its predetermined locations using traveling overmatch movement techniques. All troops will move at a fast pace to the nearest concealed position. Once at the concealed assembly point, the chalk leader will make a quick count of personnel and equipment and then proceed with the mission.

Section III

DUTIES OF KEY PERSONNEL

E-5. General.

To ensure that an air assault is executed in an effective and efficient manner, key personnel are designated to perform specific duties. This section will address the duties and responsibilities of unit leaders during air assault operations and discuss the duties and responsibilities of key personnel in the PZ control party.

E-6. Unit leader duties and responsibilities.

Platoon air assault operations.

a. **Platoon leader.** He:

(1) Has overall responsibility for the air assault operation. May act as the PZCO.

(2) Plans the operation.

(3) Briefs subordinate leaders.

(4) Issues OPORD.

(5) Conducts rehearsals.

(6) Rides in the air mission commander's aircraft to ensure better command, control, and communication.

b. **Platoon sergeant.** He:

(1) Sets up the PZ.

(a) Supervises the marking of the PZ.

(b) Supervises the clearing of obstacles from the PZ.

(2) Briefs all chalk leaders.

(3) Supervises all activity on the PZ:

(a) PZ security.

(b) Movement of troops and equipment.

(c) Placement of chalks and slingloads.

(4) Devises and disseminates the bump plan.

(5) Rides in the last aircraft for control purposes and will ensure that the PZ is cleared.

c. **Chalk leader.** He:

(1) Briefs his personnel on their respective tasks and positions inside the aircraft.

(2) Ensures that the lights or panels (if required) for his aircraft are properly emplaced.

(3) Assigns respective areas of security to his personnel. Ensures that each soldier goes to his proper area.

(4) Supervises the loading of his squad and attachments into the aircraft to ensure that all personnel assume assigned positions and have buckled their lap belts.

(5) Keeps current on location by use of his map and communication with the aircraft crew, during air movement.

(6) Ensures, upon landing, that all personnel exit the aircraft quickly, rush to a safe distance (10 to 15 meters) from the aircraft, assume the prone position, and prepare to return enemy fire.

E-7. PZ control party.

The PZ control party is responsible for the organization, control, and all coordinated operations in the PZ. Keeping in mind the unit leader's duties and responsibilities previously stated, a PZ control party for a platoon air assault operation could be organized as depicted below:

a. **PZ control officer.** He is a rifle platoon leader.

b. **PZ control noncommissioned officer in charge.** He is a platoon sergeant.

c. **RATELO, with three radios.** One radio would monitor the combat aviation net for communication with the aircraft. The second radio would be used for communication with the platoon subordinate units. The third would operate in the company command net.

d. **Chalk-linkup guides.** There is one per chalk. Their primary duties are to assist in linkup and movement of chalks from the unit assembly area to the chalk assembly areas. For platoon-size air assault operations, these guides should come from the same chalk-squad they are assigned to.

e. **Lead aircraft signalman.** He is responsible for visual landing guidance for the lead aircraft. This signalman could come from the chalk or squad loading on the lead aircraft.

f. **Slingload teams.** A team includes a signalman and hookup men (two soldiers).

Section IV

AIR ASSAULT COMMUNICATIONS

E-8. General.

The purpose of this section is to describe air assault communications as they apply to a platoon air assault operation.

E-9. Communications.

a. During establishment and operation of a platoon PZ, communications must be maintained with the aviation elements to control the aircraft, the company, and the security teams.

b. During the air movement, radio listening silence will be maintained on the company and platoon radio nets unless directed otherwise.

(1) On board the aircraft, all leaders will communicate with the troops using predetermined arm-and-hand signals or stating the message or information on a piece of paper or event map.

(2) Communication with the aircraft crew is accomplished by using the troop commander's handset.

E-10. Inflight communications.

a. To communicate with the aircraft crew, the leader will ensure the channel selector switch on the troop commander's intercommunication system (ICS) is in the "ICS" position.

b. The UH-60 also has a troop commander's antenna coax connection which is located in the aircraft. This coax enables the leader to hook up his PRC-77 radio to the FM antenna on the aircraft. This feature allows communication over the unit's command frequency. Upon landing, the RATELO unhooks the antenna coax.

E-11. Landing zone communications.

Immediately after unloading the aircraft, the RATELO will check the radios and ensure they are reconfigured as necessary. Radio communications on the LZ will be on the command frequency. Arm-and-hand signals will be utilized along with messengers to improve noise discipline.

E-12. Fire support communications.

Each rifle platoon is supported by a forward observer party. The party is equipped with a PRC-77 which is used to call for, adjust, and direct fire support. The party maintains communication with both the supporting mortars and artillery.

Section V

SLINGLOAD HOOKUP OPERATIONS

E-13. General.

Three personnel are normally used as the ground crew in slingload operations on the PZ/LZ: a signalman, a static-probe man, and a hookup man.

E-14. Static-probe equipment.

The static electricity probe consists of an insulated contact rod joined to a length of metallic tape or wire, which in turn is attached to a ground rod. The ground rod is driven into the ground and the contact rod is held to the cargo hook of the helicopter by the static-probe man.

E-15. Protective equipment.

All ground crew personnel will wear the following protective equipment:

- Helmet.
- Protective mask or dust goggles.
- Earplugs.
- Gloves (with shirt sleeves rolled down).

E-16. Ground crew emergency procedures.

In an emergency, the ground crew will move to the helicopter's right and the helicopter will move to its left.

a. The signalman will move out of the helicopter's flight path by moving to the helicopter's right.

b. The hookup team will work on the right side of the load in order to move out to the right of the aircraft in case of emergencies.

Section VI

AIR ASSAULT SAFETY BRIEFING

E-17. General.

Prior to an air assault operation, the unit chain of command will give a safety briefing to all personnel. All leaders will enforce strict safety measures when working with helicopters.

E-18. Safety briefing checklist.

Include, as a minimum, the following:

a. Identification tags and earplugs will be worn at all times when near or in an aircraft.

b. Helmets with chin straps fastened will be worn at all times.

c. **Helicopter safety measures.**

(1) UH-1Hs and UH-60s are approached from the front, forward of the rear cargo doors. **DO NOT** go near the tail of the aircraft.

(2) Approach and depart helicopters in the crouched position for extra clearance below the main rotor blade.

(3) Move to or from helicopters on sloping ground on the DOWN slope side to avoid the main rotor.

d. Shirt sleeves will be rolled down whenever working with aircraft.

e. M16 rifles will be carried with the muzzle pointed DOWNWARD, pistol grip forward, bolt closed, magazine in the weapon, and weapon on SAFE. Rounds will NOT be chambered; bayonets will not be fixed.

f. Hand grenades will be secured.

g . Short antennas will be bent completely down and long antennas will be tied down when using radios in proximity of helicopters.

h. Seatbelts are fastened upon entering the helicopter and left buckled until the crew chief signals to exit the aircraft.

i. In the event of a forced landing, all personnel will lean forward with their heads down until the aircraft comes to rest. No one will exit the aircraft until the main rotor has stopped.

APPENDIX F

Air Assault Training

Section I

SUCCESS

F-1. General.

The Army's ultimate goal is to prepare units to be able to deploy, fight, and win. The modern battlefield is more lethal than ever before and demands that every unit be combat ready in order to survive and succeed. Success depends on a unit's ability to cope with the dynamics of battle.

F-2. Ready now.

The training program must produce trained individuals in combat-effective units capable of fighting and winning. Training is a full-time job for all commanders, regardless of other operations or missions.

Section II

TRAINING

F-3. General.

Air assault training is integrated into unit programs on a routine basis to develop capability at each level from squad through battalion.

F-4. Objectives.

Commanders at all echelons are responsible for their unit's air assault training. The objective is for units to conduct air assault operations with speed, precision, and confidence. Infantry units, as well as other combat, combat support, and combat service support units, should routinely receive such training.

F-5. Small unit tactics.

Standard infantry small unit tactics and techniques are the basis for the ground phase of air assault operations. The commander ensures that all units are proficient in these tactics. He then combines this training with the other phases peculiar to air assault operations: landing, air movement, loading, and staging. He emphasizes rapid loading and unloading of aircraft, as well as quickly organizing maneuver elements **on** the landing zone to take advantage of the speed and mobility of air assault operations.

F-6. Small unit leader training.

The commander trains his small unit leaders to operate independent of their parent organization to accomplish their part of the overall mission. Additionally, small unit leaders must be able to take charge in the absence of their superiors. The speed and complex nature of air assault operations dictates the use of SOPS and battle drills.

F-7. Mobility.

The commander trains his units to travel light, consistent with the mission, taking only necessary equipment and supplies.

F-8. SOP and training.

a. Procedures for conducting air assault operations are included in unit standing operating procedures. While SOPS include routine actions that personnel might have to complete during an operation, they must also include procedures for downed aircraft, bump plans, or other conditions that may occur during the conduct of air assault operations.

b. The training Program includes sufficient training to ensure that personnel are familiar with, and proficient in, the procedures contained in the SOP.

c. During training, the information in the SOP is evaluated for completeness, simplicity, and applicability. Procedures are refined as necessary.

F-9. Land navigation.

Land navigation proficiency by all leaders is critical to success. Leaders must learn to locate positions, navigate to specific points, and use the terrain to their advantage.

F-10. Field artillery support training considerations.

a. The fire support officer and fire support team from the fire support unit train with the maneuver unit. They become familiar with the maneuver unit's SOP and teach selected personnel of maneuver units how to plan for, employ, call for, and adjust artillery and mortar fires. This makes it imperative that the habitual working relationship between the fire support and maneuver units be maintained to ensure mutual understanding of operational requirements, capabilities, and limitations. That relationship is one in which the same fire support units support a particular maneuver unit for each operation.

b. To support air assault operations, the field artillery must be proficient in slingloading operations and the planning required to execute PZ and/or LZ operations. This planning requires field artillery leaders to coordinate closely with the maneuver unit that normally controls the lift assets and the aviation units involved. Slingloading techniques require frequent training for both the hookup teams and the helicopter crews. A detailed discussion of slingloading procedures is contained in FM 55-450-1.

F-11. Aviation units.

a. Aviation unit commanders assist ground unit commanders in the development of training in the technical aspects of combined aviation and ground unit training. They also ensure that their units are technically proficient.

b. The habitual working relationship between the maneuver and aviation units is maintained whenever possible.

F-12. Infantry and aviation.

Combined infantry and aviation training is integrated into the tactical training of larger units. Ground and aviation units train together in all types of weather and reduced visibility. Both elements improve procedures to develop and refine compatible SOPs for high levels of readiness. (This is important for mechanized infantry in the heavy divisions which have inadequate organic aviation support for extensive air assault operations.)

Section III

DEVELOPING AIR ASSAULT TRAINING PROGRAMS

F-13. General.

A training program for air assault operations should include those critical, individual collective skills necessary for the successful accomplishment of the wartime mission. Unit training should be concentrated in areas where weaknesses exist.

F-14. Conduct of training.

a. Air assault training begins by familiarizing individuals in aircraft procedures to include loading and unloading, crash procedures, and aircraft safety. Proficiency in these provides a foundation for collective training of ground and aviation units.

b. Collective training should include battle drills on loading and unloading as well as organizing into combat formations on the landing zone. This will allow units to maximize the speed and mobility of air assault operations.

F-15. Use of mockups.

a. Constraints on aviation unit flying hours will limit the amount of flight time available for training. Therefore, much of the individual and small unit training will have to be accomplished using aircraft mockups. Plywood and other materials from salvage sources can be used to build the mockups. They may be man-portable and are relatively inexpensive. They can be used to train individuals on how to approach a helicopter, how to get on it, and how to get off. Air assault battle drill can be taught by using mockups. Personnel from the combat support company can be trained to load weapons, equipment, supplies, and ammunition on helicopters by practicing on mockups.

b. If the unit has a local training area of adequate size, several mockups can be used to practice battle drill to include the way the unit should offload aircraft in the landing zone. The mockups can be placed in different patterns to simulate different landing formations. Four UH-60 mockups should be sufficient for platoon training.

F-16. Familiarity with helicopters.

Commanders should be aware that many soldiers have never been near a helicopter. An attempt should be made to have at least one helicopter available early in the training cycle.

F-17. Individual and unit training.

The following subjects should be included in appropriate phases of individual and unit training:

a. **Ground units.**

(1) Subjects required to attain proficiency in ground combat skills and tactics.

(2) SOP battle drills.

(3) Physical and psychological preparedness.

(4) Methods and procedures for control and guidance of aircraft.

(5) Safety procedures in and around aircraft.

(6) Control and adjustment of supporting fires.

(7) Subjects required to attain proficiency in preparing internal and external aircraft loads.

(8) Practical experience in land and aerial navigation.

(9) Employment of attack helicopter units.

(10) Aircraft troop commander duties.

(11) Rappelling from helicopters.

(12) Downed aircraft procedures.

(13) LZ and/or PZ selection.

(14) PZ control.

(15) Proficiency in CS and CSS skills and techniques.

b. **Aviation units.**

(1) Operations planning.

(2) Nap-of-the-earth flying techniques and navigation.

(3) Formation flying.

(4) Marginal weather and reduced visibility flying techniques.

(5) Camouflage and security of aircraft.

(6) Employment of aerial weapon systems.

(7) Aircraft maintenance in a combat field environment.

(8) Unit control of aircraft and air traffic.

(9) Pathfinder procedures and techniques.

(10) Flight operations in confined areas with maximum loads.

(11) Operations with external loads.

(12) Aerial reconnaissance and security techniques.

(13) Battle drills.

c. **Subjects common to aviation and ground units.**

(1) Threat organizations and doctrine.

(2) Recognition of Threat vehicles and antiaircraft weapons and knowledge of their capabilities.

(3) Conduct of liaison and coordination.

(4) Forward refueling techniques.

(5) Training in defense against NBC weapons.

(6) Signal security, discipline, and electronic countermeasures.

(7) Aeromedical evacuation procedures.

(8) Procedures for aerial resupply.

(9) Training in air assault SOPS.

F-18. Preparation.

Training time and resources must be used efficiently. Each element of the unit should be prepared to do its part before joining support units for combined exercises. Squad and platoons should be trained in:

- Air assault battle drill.
- Preparation of internal and external loads.

F-19. Staff training.

Staffs of ground and aviation elements must be trained in planning and conducting air assault operations with emphasis on the following:

- Capabilities arid limitations of air assault operations.
- Command and staff relationships.
- Development of a plan using reverse planning sequence.
- Fire support means and control, and fire planning for air assault operations.
- Logistical procedures and requirements for air assault operations.
- Preparation of the air movement table.

APPENDIX G

Army Airspace Management

G-1. General.

a. Airspace management (ASM) in the combat zone enhances air assault operations by promoting the safe, efficient, concurrent, and flexible use of airspace. Airspace above the corps area, but below the coordinating altitude, is designated the Army airspace control authority. It is managed by maneuver unit commanders.

b. ASM is part of the Army airspace command and control (A2C2) functional mission. It identifies airspace users and coordinates, integrates, and regulates the use of airspace as defined by geographical and altitude dimensions. ASM involves planning and executing tasks that employ air assets concurrently in the Air Land Battle. ASM contributes to the overall operational plan. Maneuver unit commanders need to achieve maximum flexibility for employing organic and support assets within their airspace. To accomplish this, airspace use is standardized, restrictions are minimized, and coordination is close and continuous.

G-2. Airspace conflict resolution.

a. Normal operational planning, execution, and SOPs prevent conflicts between airspace users. Exceptions must be resolved quickly. Maneuver unit commanders from corps through battalion establish priorities for using the airspace within the Army subarea. These priorities are the guidelines used to resolve conflicts. If conflicts cannot be resolved by using these guidelines, the commander decides which user has priority.

b. Helicopters normally operate below the jointly prescribed coordinating altitude under the control of Army tactical commanders. Fixed-wing aircraft operate above the coordinating altitude and are controlled by the airspace control authority. The coordinating altitude is flexible, situation-dependent, and serves as the upper limit of the Army airspace control subarea. Coordination is -required between the airspace management facilities of each echelon to prevent unnecessary disruption of activities. This is particularly true when aircraft are required to pass through the coordinating altitude.

c. ASM rules, procedures, communication instructions, and special joint airspace requirements are included in maneuver unit SOPs and operation plans and orders. Subordinate unit commanders, as well as coordinators, controllers, and operators, can be given authority to make on-the-spot adjustments to the ASM process when unanticipated hazards are involved. The coordination of flight operations, air defense operations, and indirect artillery fires may cause conflicts in airspace use. Normally, conflicts are resolved at the lowest level possible.

G-3. Airspace restrictions.

a. Air assault operations may require airspace restrictions within the Army airspace control subarea. Requests for airspace restrictions are forwarded to the corps airspace management element (CAME) through the division airspace management element (DAME) for joint approval by the corps commander and the airspace control authority. Army approval of airspace restrictions in division airspace below the coordinating altitude can be delegated to the division commander. Commanders, however, inform the control authority of the imposed restrictions. The DAME forwards the restrictions through the CAME to the battlefield coordination element (BCE) for coordination. The information includes the time period during which the airspace restriction applies.

b. Requests for airspace restrictions include the following information:

(1) Horizontal and vertical limits of the airspace.

(2) Effective time and duration of restriction.

(3) Controlling authority of the restriction when effective.

(4) Description of the restriction specifying to whom or what it applies.

(5) Procedures for the movement of aircraft to and from adjoining airspace.

(6) Procedures for coordinating essential flight and tactical information.

(7) Warning involving flight safety hazards.

c. The division commander forwards these restriction requirements to the BCE. The approving authorities coordinate each request, considering the impact of such airspace restrictions on all airspace users. The DAME and CAME continuously monitor airspace restrictions and initiate actions to remove them when they are no longer needed.

d. For flight operations in a designated restricted area, the commander for whom the restriction was established determines when flight plans are required. Even when flight plans are not required, the appropriate ATC facility still monitors aircraft flights within the restricted area. It must be able to clear aircraft from the area to provide information for rescue purposes.

G-4. Aircraft identification.

To engage enemy aircraft, conserve air defense resources, and reduce the risk to friendly forces, air traffic identification requirements must be compatible with air defense engagement criteria. Friendly aircraft must be identified rapidly and reliably. Electronic monitoring normally fills the need, allowing flexible aircraft employment in the area of operations. When electronic identification is not possible, visual or procedural means are used.

G-5. Division airspace management and coordination.

a. **DAME functions.** The DAME in the division's main command post has staff responsibility for coordinating, integrating, and regulating the division's airspace. The DAME consists of air defense, field artillery, Army aviation, Air Force, and ATC representatives along with supporting administrative and operations personnel. DAME functions consist of future and present operational planning and coordinating to:

(1) Identify and resolve potential conflicts concerning the use of airspace by correlating ASM information.

(2) Develop and maintain the airspace utilization map.

(3) Develop, recommend, maintain, and disseminate ASM control measures and restrictions which affect the division's airspace.

(4) Maintain and disseminate information about all restricted operation areas, standard-use Army air routes, weapon-free zones, flight corridors, significant preplanned field artillery fires, nuclear strikes, close air support strikes, Air Force and Army reconnaissance missions, major air assault aviation operations, and refueling locations and status.

(5) Monitor the status of ADA and aviation; assess and advise the commander.

(6) Maintain and disseminate the status and location of navigational aids, LZs, and PZs in the area of operations (AO).

(7) Disseminate information about, and changes in, coordinating altitudes.

(8) Disseminate requirements for airfield terminal control zones with the flight coordination center (FCC) element and CAME.

(9) Disseminate requirements for flight plans, restricted areas, ADA, and aircraft weapon-free zones.

(10) Disseminate ATC and ADA procedures to be used by aviation units for deep attack operations, including return procedures.

(11) Disseminate Air Force tactical airlift airspace use and information.

(12) Disseminate selective identification features, and friend or foe identification procedures for Army aircraft.

b. **Airspace management plan.** The DAME, in coordination with the fire support element and Air Force tactical air control party, develops a division airspace management plan. It details airspace management functions and specifies SOPS, restrictive and control measures, coordination, and ATC information requirements. The plan incorporates applicable elements of the corps' and airspace control authority's airspace management and control plans. Through close coordination with the division main and rear command posts, the DAME determines which combat, CS, and CSS activities impact on effective airspace management planning.

c. **Airspace utilization annex.** The DAME maintains continuous information on airspace use and develops an airspace utilization annex to operation plans and orders. The annex with its overlay includes specific Army and joint service airspace requirements in effect for a given operation. It also outlines the commander's priority for airspace use within the division. The annex is disseminated to appropriate command posts, the FCC, and Army aviation and air defense units. The S3 of each brigade coordinates information about airspace use in his AO with the DAME. This information helps the DAME manage the division's airspace.

d. **Airspace information.** The DAME uses graphic displays that combine ground and airspace utilization information, such as air defense, aviation, ATC, and indirect fire support. Data are maintained on current and planned restrictions and special joint-use requirements. Conflicts that cannot be resolved with command guidance, orders, and SOPs are forwarded to the G3 for resolution.

e. **ADA integration.** The DAME maintains information to assist the air defense liaison officer in recommending changes in the employment of Army air defense assets. The information includes data on the air defense situation, including air defense coverage for other command post elements. Timely reports from ADA units allow the DAME to remain abreast of the air defense situation. The DAME evaluates the impact that air defense weapons control status will have on air operations and makes appropriate recommendations to the division commander. The DAME also collects NBC, weather, air threat, and other air operations information. It disseminates this information directly to appropriate airspace users and ATC facilities.

f. **Flight coordination center.**

(1) When employed in the division area, the FCC provides a communications link between terminal facilities of division landing sites, other nearby landing areas, division CPs, other FCCs, and the corps flight operations center. The FCC:

(a) Provides in-flight following of aircraft.

(b) Monitors Army aircraft operations and advises aircrews of hostile activity in the airspace.

(c) Provides information on air traffic movement within its assigned area.

(d) Passes instrument flight plans to the airspace management center for approval.

(e) Passes visual flight plans to the appropriate ATC facility.

(2) The FCC establishes liaison with the ADA command post. The ADA unit's radars receive real-time input from associated fire units. They can provide the FCC increased low altitude radar coverage over the division and forward of the FLOT by voice and data links through the ADA command post. FCC liaison with the ADA command post links Army air defense, Army aviation, and Air Force systems.

g. **Flight routes.**

(1) With the supporting ATC unit, the DAME develops an instrument flight route structure within the division area. The number of routes depends on the terrain, the number of expected flights, and the air defense threat. The division instrument route structure includes

feeder or connecting routes with adjacent divisions and corps instrument route structures. Instrument recovery routes are provided from each brigade area to facilitate recovery of aircraft which have inadvertently flown into instrument meteorological conditions. The DAME selects the location for all instrument landing sites. There are at least two precision terminal approach sites in the division. At least one of these precision sites is dedicated to logistical and medical support. Nondirectional beacon navigational aids (NAVAIDS) are also used for nonprecision terminal approaches. The ATC unit operates approach equipment, NAVAIDS, and lighting systems. The division instrument route structure, feeder routes, recovery routes, location of NAVAIDS, and instrument approach sites are included in the airspace utilization annex.

(2) In the division rear area, visual meteorological condition (VMC) flights continue under the basic guidelines and principles that apply to the battalion and brigade areas. Monitoring the division FCC frequencies, aircrews may request flight assistance including flight-following and current information on weather, NBC, airspace restrictions, and air operations.

G-6. Brigade airspace management and coordination.

a. Airspace management at brigade is primarily real-time monitoring and control of airspace and prevention of conflict between airspace users. Future operational planning is limited mostly to recommendations for control measures and restrictions necessary to support brigade and division operations. The maneuver brigade commander or his designated representative, normally the S3, is the airspace manager. The commander routinely exercises coordination through his staff, primarily the S3 assisted by the S3 Air, air liaison officer, fire support coordination officer, ADA LO, subordinate unit commanders and, when attached, the aviation LO.

b. ASM is accomplished primarily by procedural communication and visual control means. At this level, ASM functions involve detailed coordination and integration of TACAIR, indirect fire, organic air defense, and tactical fire and maneuver operations. The maneuver unit commander employs, controls, and coordinates the use of airspace by the forces supporting or reinforcing his operations. The commander also coordinates his airspace activities with other elements of the airspace control system.

c. In the brigade area, air assault operations are conducted on a "see and be seen" basis to prevent aircraft collisions. Fire support and aviation operations are conducted simultaneously. The S3 must ensure close coordination among all airspace users to prevent conflicts in airspace usage.

d. Aviation unit Operations, when possible, provide advance entry information briefings to crews entering the brigade area. These briefings include the supported unit's tactical situation. The supported units (brigades or battalions) must know in advance when and where Army and other service aircraft will enter the area. Army aircraft operating in the brigade and battalion areas are routinely controlled through the chain of command. Commanders communicate directly with the supporting or aviation unit commander to convey tasks and to coordinate missions.

G-7. Air assault task force airspace management and coordination.

a. The AATF airspace manager is the AATFC. The AATF S3 controls the use of this airspace. The capability of an AATF to conduct airspace command and control is extremely limited and consists primarily of reporting force and weapon location. The commander and S3 know the tactical plan, including proposed locations and flight routes. The operations staff elements, under supervision of the S3, collect pertinent information to keep the commander and staff informed of potential conflicts among airspace users in the AATF area. With rare exceptions, ASM at the AATF level is primarily prevention of conflict between airspace users.

b. Aircrews are briefed on tactical situations and plans by aviation unit operations personnel, or during the air mission briefing. The AATF minimizes controls, restrictions, and communications and reporting requirements. The AATF expedites the information flow and issues situation-update advisories upon request or when required.

APPENDIX H

Moving Vehicle-mounted TOW by Helicopter

Section I

UH-60

H-1. General.

Moving the TOW and its carrier about the battlefield by helicopter provides the commander with the means to rapidly emplace or shift his antiarmor assets. This appendix explains how to accomplish moves by helicopter under varying circumstances.

H-2. Transport by UH-6O.

There are a number of system combinations for consideration in moves by helicopter (Figure H-1).

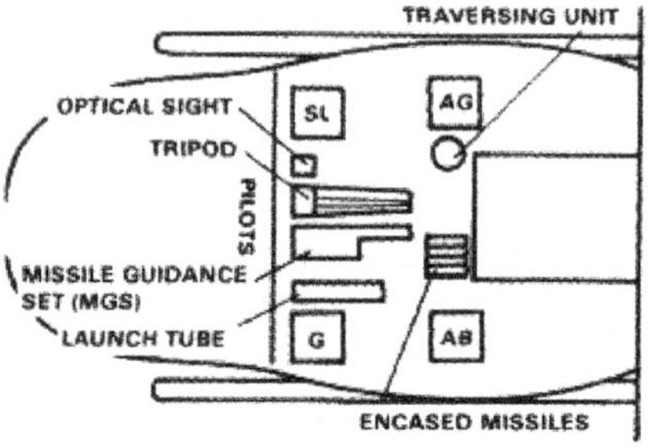

Each squad member carries the following items:

Squad leader (SL)	tripod and optical sight
Gunner (G)	missile guidance control set and launch tube.
Assistant gunner (AG)	traversing unit.
Ammunition bearer (AB)	missile (and squad radio).

Figure H-1. TOW squad located in UH-60.

a. ForunitsequippedwiththeMl5l, orM966high-mobility, multipurpose, wheeled vehicle, two UH-60 helicopters are required. In units equipped with the M151 vehicle, one helicopter carries the squad, the weapon system, and slingloads one MI 51, while the other slingloads the M 151 and trailer. In units equipped with the M966, one helicopter carries the weapon system, squad, and ammunition, while the other helicopter carries the M966 by slingload.

NOTE: Limitations imposed by conditions of density altitude may preclude the UH-60 being able to lift the M966. In this case, the squad and weapon system would be transported by the helicopter while the M966 is moved overland by the driver to link up with the rest of the squad.

b. In units equipped with the M901, or M113, the TOW system must be removed from the vehicle. The dismounted system, squad leader, gunner, and assistant gunner can then be moved by the UH-60. The M901, or Ml 13, must then be moved by the driver overland to link up with the rest of the squad.

H-3. Loading the TOW system in the UH-60.

a. The squad leader sits where he can see the LZ during approach and remain terrain-oriented. The assistant gunner sits on the same side next to the squad leader. This speeds up weapon emplacement since the squad leader carries the tripod and sight and the assistant gunner carries the traversing unit. TOW components and missiles are placed in the aircraft to be readily available to the soldier responsible for carrying them.

b. Additional missiles are carried in the aircraft and stockpiled on the landing site for the squad to retrieve. As firing positions are selected, the missiles are collected and moved to these positions. Once dismounted, the squad can move the TOW system short distances. To manually transport the TOW and several missiles for any extended distance is difficult. For this reason, the TOW system is landed and picked up as close to its firing position as possible.

Section II

CH-47

H-4. General.

Transport methods, using the CH-47, are similar to those employed with the UH-60.

H-5. Transport by CH-47.

In the event that CH-47 helicopters are available, the task is simplified since the CH-47 can lift either the squad's two M151s or one M966 and carry the squad as well. If the unit is equipped with the M901, or M113, it will have to transport the weapon system and crew with the aircraft and have the carrier follow overland.

NOTE: For information on how to prepare vehicles for either internal load or external load, see FM 57-38.

Glossary

Acronyms and Abbreviations

A

A^2C^2 Army airspace command and control

AATF air assault task force

AATFC air assault task force commander

ABN air battle net

ACA airspace coordination area

ACL allowable cargo load

ACP air control point

ACS air cavalry squadron

AD air defense

ADA air defense artillery

ADAM area denial artillery munition

admin-log administrative-logistics

AGL above ground level

AHB attack helicopter battalion

ALO air liaison officer

altn alternate

am amplitude modulation

AMB air mission briefing

AMC air mission commander

ammo ammunition

ANGLICO air and naval gunfire liaison company

AO area of operations

ASH advanced scout helicopter

ASK airspace management

AT antitank

ATC air traffic control

atk attack

AVN aviation

B

BCE battlefield coordination element

Bde brigade

Bn battalion

BSA brigade support area

BTC battle team captain

C

C^2 command and control

C^3 command, control, communications

CAB combat aviation brigade

CAC combat aviation company

CAME corps airspace management element

CAN combat aviation net

CAS close air support

CAV cavalry

CCP communications checkpoint

cdr commander

CEO communication-electronics officer

CEOi communications-electronics operation instructions

CFA covering force area

CH cargo helicopter

CL chalk leader

CO commanding officer

co company

corr corridor (flight)

CP command post

Cs combat support

csc combat support company

css combat service support

D

DAME division airspace management element

DPICM dual-purpose, improved conventional munitions

DS direct support

E

ea each

ECM electronic countermeasures

EEI essential elements of information

EMP electromagnetic pulse

EN engineer

EPW enemy prisoner of war

equip equipment

EW electronic warfare

F

FA field artillery

FAC forward air controller

FARP forward rearming and refueling point

FASCAM family of scatterable mines

FAST forward area support team

FCC flight coordination center

FDC fire direction center

FEBA forward edge of the battle area

FIST fire support team

FLOT forward line of own troops

fit flight

FM frequency modulation

FO forward observer

FRAGO fragmentary order

FSB forward support battalion

FSCOORD fire support coordinator

FSE fire support element

FSO fire support officer

FSSE forward service support element

G

GS general support

GSR ground surveillance radar

H

HC chemical smoke

HE high explosive

hel bn helicopter battalion

HF high frequency

HQ headquarters

hr hour

ht height

HVT high value targets

I

ICM improved conventional munitions

ICS (commander's) intercommunication system (aircraft)

ILLUM illumination

INF infantry

INTSUM intelligence summary

IPB intelligence preparation of the battlefield

J

JAAT joint air attack team

K

km kilometer

L

LAPES low altitude parachute extraction system

lb pound

LC line of contact

Ldr leader

LO liaison officer

LOGPAC logistics package

LRSU long-range surveillance unit

LZ landing zone

M

MANPAD man-portable air defense (system)

max maximum

MBA main battle area

medevac medical evacuatio

METT-T mission, enemy, terrain, troops, time available

ml military intelligence

min minute

MOPP mission-oriented protection posture

mort mortar

N

NATO North Atlantic Treaty Organization

NAVAIDS navigational aids

NBC nuclear, biological, chemical

NCO noncommissioned officer

NCOIC noncommissioned officer in charge

NCS net control station

NGF naval gunfire

NLT not later than

NOE nap of the earth (flight)

nuc nuclear

NVG night vision goggle

0

obi objective

OCOKA Observation and fields of fire, Cover and concealment, Obstacles and movement, Key terrain, Avenues of approach

OH observation helicopter

OP observation post

OPCON operational control

OPORD operation order

OPSCD operations schedule

OPSEC operations security

P

pax passenger

PLL prescribed load list

plt platoon

POL petroleum, oils, lubricants

prep preparation

prim primary

PSG platoon sergeant

PW prisoner of war

PZ pickup zone

PZCO pickup zone control officer

R

RAAMS remote antiarmor mine system

RAP rocket assisted projectile

RATELO radiotelephone operator

RATT radio teletypewriter

recon reconnaissance

rd round

RFA restrictive fire area

RFL restrictive fire line

RP release point

RRP replacement receiving point

rt route

rtn return

S

SEAD suppression of enemy air defense

sec section

SHORAD short-range air defense

SIGSEC signal security

SLAR side-looking airborne radar

SOP standing operating procedure

S/L staggered trail left (formation)

SP start point

S/R staggered trail right (formation)

SSB single sideband (radio)

T

TACAIR tactical air

TAC CP tactical command post

TACP tactical air control party

TACSATCOM tactical satellite communications

TF task force

tgt target

tm team

TOC tactical operations center

TOW tube-launched, optically tracked, wire-guided

U

UH utility helicopter

UHF ultra high frequency

USAF United States Air Force

USMC United States Marine Corps

V

VHF very high frequency

Vic vicinity

VMC visual meteorological conditions

W

WP white phosphorus

Wt weight

X

XO executive officer

References

Required Publications

Required publications are sources that users must read in order to understand or to comply with this publication.

Field Manuals (FMs)

FM 1-100 Combat Aviation Operations

FM 7-30 Infantry, Airborne, and Air Assault Brigade Operations

Related Publications

Related publications are sources of additional information. Users do not have to read them in order to understand this publication.

Army Regulations (ARs)

AR 95-Series Army Aviation Regulations

AR 310-25 Dictionary of United States Army Terms

AR 310-50 Authorized Abbreviations, Brevity Codes, and Acronyms

AR 350-1 Army Training

AR 525-Series Military Operations Regulations

Field Manuals (FMs)

FM 1-101 Aircraft Battlefield Countermeasures and Survivability

FM 1-103 Airspace Management and Army Air Traffic in a Combat Zone

FM 1-104 Forward Arming and Refueling Points

FM 1-107 Air-to-Air Combat

FM 1-202 Environmental Flight

FM 1-402 Aviator's Recognition Manual

FM 3-3 NBC Contamination Avoidance

FM 3-4 NBC Protection

FM 3-5 NBC Decontamination

FM 3-100 NBC Operations

FM 5-100 Engineer Combat Operations

FM 5-101 Mobility

FM 5-102 Countermobility

FM 5-103 Survivability

FM 6-20 Fire Support in Combined Arms Operations

FM 7-10 The Infantry Rifle Company (Infantry, Airborne, Air Assault, Ranger)

FM 7-20 The Infantry Battalion (Infantry, Airborne and Air Assault)

FM 8-35 Evacuation of the Sick and Wounded

FM 11-50 Combat Communications Within the Division

FM 17-47 Air Cavalry Combat Brigade

FM 17-50 Attack Helicopter Operations

FM 17-95 Cavalry

FM 19-4 Military Police Team, Squad, Platoon Combat Operations

FM 21-6 How to Prepare and Conduct Military Training

FM 21-60 Visual Signals

FM 21-76 Survival, Evasion, and Escap

FM 24-1 Combat Communications

FM 30-5 Combat Intelligence

FM 44-1 US Army Air Defense Artillery Employment

FM 44-3 Air Defense Artillery Employment, Chaparral/Vulcan/Stinger

FM 44-8 Small Unit Self-Defense Against Air Attack

FM 55-9 Unit Air Movement Planning

FM 55-12 Movement of Army Units in Air Force Aircraft (AFM 76-6)

FM 55-450-1 Army Helicopter External Load Operations

FM 55-450-2 Army Helicopter Internal Load Operations

FM 57-38 Pathfinder Operations

FM 71-101 Infantry, Airborne, and Air Assault Division Operations

FM 90-1 Employment of Army Aviation Units in a High Threat Environment

FM 90-13 River Crossing Operations

FM 90-14 Rear Battle

FM 100-5 Operations

FM 100-10 Combat Service Support

FM 100-27 US Army/US Air Force Doctrine for Joint Airborne and Tactical Airlift Operations

FM 101-5 Staff Organization and Operation

FM 101-5-1 Operational Terms and Symbols

FM 101-10-1 Staff Officers'Field Manual: Organizational, Technical, and Logistical Data (Unclassified Data)

Training Circular (TC)

TC 1-65 Helicopter Operations from Amphibious Assault Ships

Army Training and Evaluation Program (ARTEP)

ARTEP 7-15-MTP Infantry Battalions (Infantry, Airborne, Air Assault, Ranger)

Department of the Army Pamphlets (DA Pams)

DA Pam 310-1 Consolidated Index of Army Publications and Blank Forms

DA Pam 310-35 Index of International Standardization Agreements

TRADOC Training Text (TT)

TT 17-50-3 Joint Air Attack Team (JAAT) Operations

International Agreements*

The provisions of this manual are the subject of the international agreements listed.**

Title	NATO STANAG No.	QSTAG No.
Operation Orders, Annexes to Operation **Orders,** and Administrative and Logistic Orders	2014	506
Relief of Combat Troops	2082	
Principles and Procedures for Establishing Liaison	2101	533
Procedures for Marshalling Helicopters in Airmobile Operations	2351	585

Procedures for the Employment of Helicopters in the Antiarmor Role in Support of Ground Forces	2355	277
Principles of Engagement for the Landing Sites	2860	
Procedures for the Recovery of Downed Aircraft/Helicopters While Engaged in Airmobile Operations	2861	586
Minimum Navigational Facilities for Multi-National Airmobile Operations	2863	

***NOTE:** Standardization Agreements available from:

Naval Publications and Forms Center (NPFC)
5801 Tabor Avenue
Philadelphia, PA 19120

(DD Form 1425 can be utilized to requisition documents).

**The acronyms are defined as follows:
North Atlantic Treaty Organization (NATO)
Standardization Agreement (STANAG)
Quadripartite Standardization Agreement (QSTAG)

Title	**NATO STANAG No.**	**QSTAG No.**
Planning and Coordination Procedures for Airmobile Operations	2876	
Safety Regulations for Helicopter Users	2878	44
Airmobile Operations (ATP 41)	2904	665
Helicopter Tasking Messages	2956	
Aircraft Marshalling Signals	3117	585
Forward Air Controllers Supplement No. 1 to ATP-27(A)	3325	
Data Forms for Planning Air Movements	3345	44
General Rules Covering the Transport of Loads by Helicopters	3468	328
Transport of Troops by Helicopter	3532	570
Drop Zones and Extraction Zones	3570	44
Helicopter Tactical or Non-Permanent Landing Sites	3597	44

Helipad Markings	3619	
Helicopter Day and Night Formation Flying	3627	44
Helicopter Tactical Refueling	3628	44
Helicopter Tactical Operations at the High Hover	3630	570
Terrain Flight with Helicopters	3782	
Doctrine and Procedures for Airspace Control in the Combat Zone	3805	45
Procedures for Establishing Communications	5048	522

Chart D-1. Army aircraft characteristics.

A AIRCRAFT	OH-6A	OH-58A	OH-58C	CH-47A	CH-47B	CH-47C	CH-47D
B NORMAN CREW	1+(OBS)	1+(OBS)	1+(OBS)	4	4	4	4
C OPERATIONAL CHARACTERISTICS 2,3,4 (1) MAX ALLOWABLE GROSS WEIGHT	2400	3000	3200	33,000	40,000	46,000	50,000
(2) BASIC WEIGHT	1163	1586	1896	18,153	19,591	20,481	22,499
(3) USEFUL LOAD	1237	1417	1302	14,888	20,445	23,380	27,501
(4) PAYLOAD/NORMAL MISSION	650 [8]	760 [8]	837 [8]	10,000	15,000	18,200	20,206
(5) FUEL CAPACITY [9]	400/615	475/73	465/715	4036/621	7351/1131	6695/1030	8794/1353
(6) FUEL CONSUMPTION RATE [9]	143/22	189/29	175/27	2120/342	2780/427	3038/467	2600/400
(7) NORMAL CRUISE SPEED	121	120	120	120	150	155	155
(8) ENDURANCE AT CRUISE (PLUS 30 MIN RESERVE)	3+15	3+30	3+00	1+30	1+00	2+00	2+30
(9) GRADE OF FUEL	JP-4	JP-4	JP-4	JP-4	JP-4	JP-4	JP-4
D PASSENGER CAPACITY (1) TROOP SEATS	3	4	4	33	33	33	33
(2) NORMAL CAPACITY	3	4	4	33	33	33	33
(3) TOTAL CAPACITY W/CREW	4	4	4	37	37	37	37
(4) LITTERS & AMBULATORY	N/A	2/4	2	24	24	24	24
E EXTERNAL CARGO (1) MAXIMUM RECOMMENDED EXTERNAL LOAD [5]	N/A	N/A	N/A	16,000	20,000	20,000	28,000
(2) RESCUE HOIST CAPACITY	N/A	N/A	N/A	600	600	600	600
(3) CARGO WINCH CAPACITY	N/A	N/A	N/A	3000	3000	3000	3000
F DIMENSIONS (1) LENGTH - FUSELAGE [6]	23'-0"	32'-3.5"	32'-8.8"	51'-0"	51'-0"	51'-0"	51'-0"
(2) LENGTH - BLADES UNFOLDED	30'-4"	40'-11.8"	40'-11.8"	98'-3"	99'-0"	99'-0"	99'-0"
(3) LENGTH - BLADES FOLDED	23'-0"	N/A	N/A	51'-0"	51'-0"	51'-0"	51'-0"
(4) WIDTH - BLADES FOLDED	5'-6"	N/A	N/A	12'-5"	12'-5"	12'-5"	12'-5"
(5) WIDTH - TREAD	6'-9"	6'-3.5"	6'-5.4"	11'-11"	11'-11"	11'-11"	11'-11'
(6) HEIGHT - EXTREME	8'-3"	9'-6.5"	12'-0"	18'-6"	18'-8"	18'-8"	18'-8"
(7) DIAMETER - MAIN OR FORWARD ROTOR	24'-4"	35'-4"	35'-4"	59'-1"	60'-0"	60'-0"	60'-0"

(8) DIA. TAIL OR REAR ROTOR	4'-3"	5'-2"	59'-1"	60'-0"	60'-0"	60'0"	60'-0"
(9) WING SPAN	N/A	N/A	N/A	N/A	N/A	N/A	N/A
G CARGO DOOR (1) DIMENSIONS - WIDTH HEIGHT	41x34.5	40x35	40x35	90x78	90x78	90x78	90x78
(2) LOCATION - SIDE OF FUSELAGE	LEFT & RIGHT	LEFT & RIGHT	LEFT & RIGHT	REAR	REAR	REAR	REAR
H CARGO COMPARTMENT (1) FLOOR - ABOVE GROUND	14.5"	22.5"	22.5"	30"	31.2"	31.2"	31.2"
(2) USABLE LENGTH	69"	39"	39"	360"	360"	362"	362"
(3) FLOOR WIDTH	38"	50"	50"	90"	90"	90"	90"
(4) USABLE HEIGHT	38"	50"	50"	78"	78"	78"	78"
(5) MAXIMUM CARGO SPACE	40	20	20	1474	1474	1474	1474
I WEAPONS [10]	SM-27E-1	SM-27E-1	N/A	M-24 XM-41	M-24 XM-41	M-24 XM-1	M-24 XM-41

A AIRCRAFT	CH-54A	CH-54B	UH-1C/M	UH-1H/V	UH-60	TH/AH-1G	AH-1S	AH-64 [14/23]
B NORMAL CREW	4	4	2	2	3	2	2	2
C OPERATIONAL CHARACTERISTICS (1) MAX ALLOWABLE GROSS WEIGHT	42,000	47,000	9500	9500 .	20,250	9500	10,000	17,400
(2) BASIC WEIGHT	20,800	21,200	4827	5132	10,500	5560	6598	10,505
(3) USEFUL LOAD	21,200	25,800	4673	4368	6195	3940	3402	6895
(4) PAYLOAD/NORMAL MISSION	11,650 [8]	16,250 [8]	2685	2900	3360 [17]	1785 [21]	1293 [21]	4090 [21]
(5) FUEL CAPACITY [9]	8794/1353	8794/1353	1573/242	1358/209	2360/362	1755/270	1703/262	2405/370
(6) FUEL CONSUMPTION RATE [9]	3614/556	4230/651	500/77 550/84 [22]	550/84	960/148	546/83.6	640/98	810/124
(7) NORMAL CRUISE SPEED	95	10	92-140	90-120	145	0-190 [20]	0-190 [20]	0-161 [20]
(8) ENDURANCE AT CRUISE (PLUS 30 MIM RESERVE)	2+00	1+30	3+00-2+45	2+15	2+15	2+00 [7]	2+30	1+45
(9) GRADE OF FUEL	JP-4/5	JP-4/5	JP-4/5	JP-4/5	JP-4/5/8	JP-4	JP-4	JP-4/5/8
D PASSENGER CAPACITY (1) TROOP SEATS	1	1	7	11	14	0	0	0
(2) NORMAL CAPACITY	1	1	7	11	14	0	0	0
(3) TOTAL CAPACITY W/CREW	5	5	9	13	17	0	0	0
(4) LITTERS & AMBULATORY	0	0	3	6	4/6	0	0	0
E EXTERNAL CARGO (1) MAXIMUM RECOMMENDED EXTERNAL LOAD [5]	20,000	25,000	3787	4000	8000	N/A	1380 [19]	6200 [19]

(2) RESCUE HOIST CAPACITY	N/A	N/A	300 [18]	300 [18]	600	N/A	N/A	N/A
(3) CARGO WINCH CAPACITY	15,000	25,000	N/A	N/A	N/A	N/A	N/A	N/A
F DIMENSIONS (1) LENGTH - FUSELAGE [6]	70'-0"	70'-0"	42'-7"	40'-7"	50'-7.5"	44'-5.2"	44'-7"	49'-3"
(2) LENGTH - BLADES UNFOLDED	88'-5"	88'-5"	52'-10"	57'-1"	64'-10"	52'-11.7"	53'-1"	57'-1"
(3) LENGTH - BLADES FOLDED	N/A	N/A	N/A	N/A	40'-4'	N/A	N/A	N/A
(4) WIDTH - BLADES FOLDED	N/A	N/A	N/A	8'-7'	9'-8.1"	10'-4"	10'-9"	16'-3"
(5) WIDTH - TREAD	19'-9"	19'-9"	8'-4"	8'-7"	8'-10.2"	7'-0"	7'-0"	6'-6"
(6) HEIGHT - EXTREME	24'-5"	24'-5"	12'-8"	14'-6"	17'-6"	11'-7"	13'-9"	12'-6"
(7) DIAMETER - MAIN OR FORWARD ROTOR	72'-0"	72'-0"	44'-0"	48'-0"	53'-8"	44'-0"	44'-0"	48-0'"
(8) DIA. TAIL OF REAR ROTOR	16'0"	8'-6"	8'-6"	11'-0"	8'-6"	8'-6"	9'-3"	
(9) WING SPAN	N/A	N/A	N/A	N/A	N/A	10'-4"	10'-4"	16'-3"
G CARGO DOOR (1) DIMENSIONS - WIDTH HEIGHT	104.5 (POD)	104.5 (POD)	48x48	74x48	68x54	N/A	N/A	N/A
(2) LOCATION - SIDE OF FUSELAGE	REAR	REAR	LEFT & RIGHT	LEFT & RIGHT	LEFT & RIGHT	N/A	N/A	N/A
H CARGO COMPARTMENT (1) FLOOR - ABOVE GROUND	27" (POD)	27" (POD)	14"	24"	19"	N/A	N/A	N/A
(2) USABLE LENGTH	329"	329"	60"	92"	110"	N/A	N/A	N/A
(3) FLOOR WIDTH	104.5"	104.5"	80.5"	96"	72"	N/A	N/A	N/A
(4) USABLE HEIGHT	78"	78"	54"	49"	54"	N/A	N/A	N/A
(5) MAXIMUM CARGO SPACE	1552	1552	140	220	246.8	N/A	N/A	N/A
I WEAPONS	N/A	N/A	XM-3 M-5 M-6 XM-16 XM-21 M-22 XM-156	M-23 M-56 M-59 [23]	M-23	M-18 M-26 M-35 M-157 [11] M-158 [11] M-159 [12] M-200 [12]	M-65 M-97 M-158 M-200 [12] M-260 M-261	XM-430 HELLFIRE M-200 [12] M-260 M-261 M-230

NOTE: This chart is for general reference only. Refer to the appropriate operator's manual.

1 A-Attack, C-Cargo, O-Observation, U-Utility

2 All data computed at standard conditions at sea level.

3 Detailed weight computations and characteristics taken from current 55 series TMs.

4 Data subject to change due to developmental testing.

5 Maximum load an aircraft is capable of lifting.

6 Dimensions from nose to end of tail.

7 Varies with load carried. Figure given is for normal mission profile.

8 Does not meet 200 NM range requirement of normal mission definition.

9 Aviation gas figured on 6 lbs/gal. JP-4 computed on 6.5 lbs/gal.

10 Indicates type of weapons aircraft can carry. Specific armament based on unit assignment.

11 7-round 2.75-inch rocket pod.

12 19-round 2.75-inch rocket pod.

13 (Deleted)

14 Subject to final develop configuration.

15 Without external fuel.

16 With external fuel.

17 Normal mission, internal load, probability exists to cube out before weight out. Max load on the floor is 300 lbs/sq ft.

18 UH-1 is restricted to hoist capacity of 300 lbs because of CG conditions.

19 External wing stores.

20 Due to armament configurations and flight profiles.

21 Considers gross weight minus basic weight minus 400 lbs for crew and total fuel weight.

22 Fuel consumption at 92 kts, 77 gal/hr; at 140 kts, 84 gal/hr.

23 Weapons are not applicable to UH-1V, MEDEVAC helicopters.

DEFINITIONS

Maximum Allowable Gross Weight. The maximum allowed total weight of the aircraft prior to takeoff. The "basic weight" of the aircraft plus the crew, personnel equipment, special devices, passengers and cargo, and usable fuel and oil. This is limited by structure, power available, or landing load.

Basic Weight. The empty weight of an aircraft in its basic configuration, to include all appointments, integral equipment, instrumentation, and trapped fuel and oil, but excluding passengers, cargo, crew, fuel, and oil.

Useful Load. The load-carrying capability of an aircraft, including payload, crew, oil, and usable fuel required for the mission. This is the difference between "maximum allowable gross weight" and "basic weight" as defined above. Thus, a reduction of the fuel load will reduce the endurance and increase payload. Full oil is required for all missions.

Payload. The useful load less the crew, full oil, and the required fuel for the mission.

Normal Mission. Payload available computed under following conditions:

- Fuel for 200 NM range plus 30-minute reserve.
- Flight altitude 2,000 feet MSL, standard temperature.
- Takeoff maximum gross weight (weight of crew included).

Normal Cruising Speed. The true airspeed which an aircraft can normally be expected to maintain at some standard power setting below rated military power. This speed will vary with altitude (for example for the U-8F, normal is 165 at 65 percent power at 8,000 feet).

Made in the USA
Monee, IL
04 November 2026

67377584R00125